SAM1R06d

REMAINING ASSET LIFE: A STATE OF THE ART REVIEW

STRATEGIC ASSET MANAGEMENT AND COMMUNICATION

by:

David Marlow

Paul Davis

Dung Trans

David Beale

Stewart Burn
CSIRO

Anthony Urquhart
MWH

2009

WERF

The Water Environment Research Foundation, a not-for-profit organization, funds and manages water quality research for its subscribers through a diverse public-private partnership between municipal utilities, corporations, academia, industry, and the federal government. WERF subscribers include municipal and regional water and wastewater utilities, industrial corporations, environmental engineering firms, and others that share a commitment to cost-effective water quality solutions. WERF is dedicated to advancing science and technology addressing water quality issues as they impact water resources, the atmosphere, the lands, and quality of life.

For more information, contact:
Water Environment Research Foundation
635 Slaters Lane, Suite 300
Alexandria, VA 22314-1177
Tel: (703) 684-2470
Fax: (703) 299-0742
www.werf.org
werf@werf.org

This report was co-published by the following organizations. For non-subscriber sales information, contact:

IWA Publishing
Alliance House, 12 Caxton Street
London SW1H 0QS, United Kingdom
Tel: +44 (0) 20 7654 5500
Fax: +44 (0) 20 7654 5555
www.iwapublishing.com
publications@iwap.co.uk

Library of Congress Catalog Card Number: 2009928931
IWAP ISBN: 978-184339-330-6/1-84339-330-1

CSIRO and MWH

The research on which this report is based was developed, in part, by the United States Environmental Protection Agency (EPA) through Cooperative Agreement No. CR-83155901 with the Water Environment Research

Foundation (WERF). However, the views expressed in this document are solely those of CSIRO and MWH and neither EPA nor WERF endorses any products or commercial services mentioned in this publication. This report is a publication of WERF, not EPA. Funds awarded under the Cooperative Agreement cited above were not used for editorial services, reproduction, printing, or distribution.

ACKNOWLEDGMENTS

The research team would like to acknowledge the kind assistance of all those who contributed to this project, especially the case study partners United Utilities, UK, Sydney Water, Australia, and The Department of Utility Services of the City of Henderson, U.S.

Research Team

Principle Investigators:
David Marlow, Ph.D.
CSIRO

Anthony Urquhart, CP Eng,
MWH

Report Production:
Paul Davis, Ph.D.
Dung Trans, Ph.D.
David Beale, Ph.D.
Professor Stewart Burn
CSIRO

WERF Issue Area Team Members (IAT)

Terry Martin, *Seattle Public Utilities*

Kevin Hadden, *Orange County Sanitation District*

Greg Cawston, *Sydney Water Corporation*

Greg Kane, *Sydney Water Corporation*

Manju Sharma, *MWRDGC*

Wayne Green, *Green Management Ltd.*

Kendall Jacob, P.E., *Cobb County Water System*

Tim Madhanagopal, *Orange County Utilities Department*

Ted Regan, *Massachusetts Water Resources Authority (MWRA)*

Steve Krai, *Los Angeles County Sanitation District*

Maureen Hodgins, *Awwa Research Foundation*

Chris Royce, *Anglian Water Services, Ltd.*

Daniel Murray, *U.S. Environmental Protection Agency*

Stephen Whipp, *United Utilities North West (UUNW)*

Steve Kaye, *Anglian Water Services, Ltd.*

Bud Templin, *Westin Engineering, Inc.*

James Thomson, P.E., *Jason Consultants LLC, Inc.*

Water Environment Research Foundation Staff

Director of Research: Daniel M. Woltering, Ph.D.

Program Director: Walter L. Graf, Jr.

ABSTRACT AND BENEFITS

Abstract:

In 2006, WERF initiated a program of research to assist water and wastewater utilities to improve strategic management of their pipelines, treatment works, and other physical assets. This report is an output of the fourth research track (Track 4) of WERF's research program 'Strategic Asset Management Communication and Implementation' (SAM1R06).The goal of Track 4 is to contribute to the development of techniques, tools and methods for estimating residual life of wastewater assets. This report provides a synthesis of knowledge on available methods to estimate "end of life" and "remaining asset life."

Drawing on the literature and the knowledge of the research team and industry partners, information is presented on the factors that influence the life of the different asset classes involved in the provision of wastewater services. A taxonomy of asset life is given, along with a critical review of the conceptual linkages between risk, asset management, and remaining asset life. A review of techniques used to assess remaining asset life is included, as well as a detailed 'state of the art' review of modeling tools and approaches.

One of the key questions to be addressed in this initial stage of the research was the state of knowledge for estimating and predicting remaining asset life, and the capacity to translate between condition and performance data (e.g., the presence of significant defects) and the residual life of an asset. In this regard, this report builds on previous work undertaken by the research team into protocols for condition and performance assessments, as detailed in WERF (2007).

Benefits:

This project summarizes the state of the art for assessing remaining asset life as it applies to wastewater assets and asset management, and the report delivers the following benefits:

- Presents a taxonomy of relevant terms and a detailed list of factors that effect remaining life for the different asset classes.
- Reviews how concepts of remaining asset life map onto asset management, including case studies that show how different water authorities address remaining asset life in their asset management strategies and approaches.
- Illustrates how risk concepts influence remaining asset life.
- Reviews the role of asset assessments in understanding remaining asset life
- Reviews the state of the art in modeling remaining asset life.
- Provides a conceptual mapping for how water authorities could apply a spectrum of modeling tools as their asset management capacity develops.

Keywords: Strategic asset management, wastewater assets, remaining asset life, utility infrastructure, risk management, asset end of life.

TABLE OF CONTENTS

LIST OF TABLES

LIST OF FIGURES

LIST OF ACRONYMS

3D	Three-Dimensional
AC	Asbestos Cement
AMP	Asset Management Plan
ANN	Artificial Neural Networks
AwwaRF	American Water Works Association Research Foundation (now WRF)
C&B	Civil and Building (asset category)
CCTV	Closed Circuit Television
CMMS	Computerized Maintenance Management System
CSIRO	Australian Commonwealth Scientific and Industrial Research Organisation
DUS	Department of Utility Services of the City of Henderson, Nevada
FORM	First-Order Reliability Method
FOSM	First-Order Second Moment
GIS	Geographic Information System
GLM	Generalized linear model
GRP	Glass reinforced plastics
ICA	Instrumentation, Control and Automation (asset category)
ICGs	Internal Condition Grades
IPART	The Independent Pricing and Regulatory Tribunal
KPI	Key Performance Indicator
M&E	Mechanical and Electrical (asset category)
MC	Monte Carlo
MTBF	Mean time between failures
NSW	New South Wales, Australia
NRC	National Research Council of Canada
O&M	Operations and Maintenance
Ofwat	Office of Water Services
PE	Polyethylene
PVC	Polyvinyl Chloride
RCM	Reliability Centered Maintenance
SAM	Strategic Asset Management
SCADA	Supervisory Control And Data Acquisition
SIMPLE	Sustainable Infrastructure Management Program Learning Environment
SRM	Sewer Rehabilitation Manual
SPG	Structural Performance Grade
SW	Sydney Water Corporation, Australia
TAM	Tactical Asset Management
TFN	Triangular fuzzy numbers
UU	United Utilities, UK
U.S. EPA	United States Environmental Protection Agency
WERF	Water Environment Research Foundation
WRF	Water Research Foundation (formerly AwwaRF)
WRc	Water Research Centre

EXECUTIVE SUMMARY

Utilities throughout the world are faced with the challenge of managing their water and wastewater assets to meet stakeholder and customer levels of service with limited funds. In the United States, the U.S. Environmental Protection Agency released a study in 2002 that highlighted the funding gap between actual and needed investment in water and wastewater infrastructure (U.S. EPA, 2002), estimated at $122 billion for wastewater and $102 billion for drinking water under current revenue scenarios. The projected funding gap is a result of aging infrastructure and increasing population and quality demands on water systems. Closing the gap likely will require both increases in funding and more efficient management of water and wastewater assets, e.g., through application of Strategic Asset Management (SAM) approaches.

In response to the needs of its utility members, the Water Environment Research Foundation (WERF) initiated a research program in 2006 to develop guidance and decision support tools for communication and implementation of SAM. The research effort is composed of four separate, but inter-related tracks:

Track 1: Public Communication to build support for SAM implementation

Track 2: Benchmarking and Case Studies of SAM programs

Track 3: Decision Analysis and Implementation Guidance for SAM

Track 4: Remaining Asset Life estimation techniques

This report is an output of Track 4 of the WERF SAM research program, with the objective to further develop the techniques, tools, and methods for estimating performance and residual economic life of assets. This research was undertaken in a staged manner, in order to provide a stepwise development of concepts and protocols. This report is the output from the first stage of the research, which had the goal of conceptually mapping out the broad range of factors that influence remaining asset life, as well as to provide a state of the art review of the science and practice of 'end of life asset management prediction'.

Remaining asset life is important because utilities throughout the world are faced with the challenge of how best to manage their existing asset stock to provide satisfactory customer service with limited funds. Effective asset management will help utilities meet this challenge, and an important element of this is the ability to assess and/or predict the remaining life of an asset. Asset life and remaining asset life are, however, multifaceted concepts. At a basic level, asset life relates to the repairability of an asset, but is also dependent on a range of other factors. Ultimately, remaining life reduces to a question of economics; assets should be either replaced or retained because it is more economic to do so. In practice, asset life is modified and extended through maintenance and capital investment.

A definition of asset life is proposed in Chapter 2.0 that links the end of asset life to the time at which a significant capital (rather than operational) intervention is required. Remaining life is then the time left before a significant capital intervention is required. This definition aligns with the needs of asset managers, who must determine when an asset is reaching its end of life because investment is then necessary and must be catered for (unless the asset is no longer needed, in which case the lack of need drives the consideration of end of asset life).

A broad range of asset-related factors influence the life of various asset classes at different stages of the asset life cycle. In addition, some factors that influence asset life are not related to the asset themselves, but derive from the economic aspects of the water authority and/or level of service the authority has to deliver to customers, the community and the environment. Chapter 3.0 provides a comprehensive listing of the factors that influence asset life, and thus remaining life, for the full range of asset classes.

Management of risk is a critical component of effective asset management and therefore risk is an important determinant of remaining asset life; Chapter 4.0 discusses asset-related risk and the linkage to remaining asset life concepts. Risk is considered to be the product of the likelihood and consequence of failure. Likelihood of failure for an asset can be conceptualized in terms of a load-capacity relationship. Final failure is often associated with a particular over-loading event, and failure probability can be considered in terms of the probability that the over-loading event will occur. Many failure consequences can be reduced to financial costs, though others are intangible.

The potential consequences of asset failure can be used to give a measure of asset importance, which in turn can be used to specify the management strategy adopted for the asset, including the time at which the asset is considered to be at the end of its life. In particular, reactively managed assets will tend to have end of life dictated by failure events, while proactively managed assets will have end of life dictated by additional measures of asset deterioration such as condition or performance.

Risk can be considered in dollar terms and therefore treated as a cost in the same way as other capital and operational expenditure items. Techniques like Life Cycle Costing or similar can be used to determine if it is more economic to retain an existing asset or renew it in some way. In theory, the optimal time for replacement is when the total expected cost is minimized. In practice, however, managers may attempt to maximize the likelihood of successful future performance at reasonable cost, rather than to minimize cost explicitly.

The practical application of risk concepts within a U.S. water authority are presented through a case study developed with The Department of Utility Services of the City of Henderson, Nevada.

In addition to risk concepts, this report gives a broad review of the role of remaining life in asset management. As discussed in Chapter 5.0, asset management is undertaken at a range of scales, both spatial and temporal. The different scales of asset management are referred to herein as 'strategic asset management', 'tactical asset management', and 'operations and maintenance'. The concept of remaining asset life plays a role in each of these aspects of asset management. For example, in strategic asset management, remaining asset life concepts are used in budget setting and asset valuation, whereas in tactical asset management the concepts are used in the justification and prioritization of planned replacement work.

The concepts of remaining asset life are illustrated through two case studies developed with United Utilities (England) and the Sydney Water Corporation (Australia).

There are different ways in which remaining life of wastewater assets can be assessed using data on the current state of an asset, and these are discussed in Chapter 6.0. The simplest approach is to use the age of the asset in combination with an estimate of the expected service life. Such approaches are, however, considered simplistic. For a broad range of assets in the wastewater and other sectors, it is possible to undertake assessments of asset condition and performance to determine remaining asset life. Interpretation of data on condition and performance can be undertaken using a range of approaches. For example, for some assets it is possible to specify a critical defect that represents the limit of acceptable condition. In addition, performance thresholds can be specified that indicate the need for some kind of intervention. The issue of determining acceptable asset state is then reduced to the need to assess whether such critical defects exist or performance thresholds have been exceeded, or to make an assessment of the time until this will occur. The concept of remaining life is linked with this latter assessment.

Wastewater management is asset intensive, and those assets tend to have a long life. Thus, effective strategic and tactical planning requires a capacity, not only to assess the remaining life of an asset or group of assets, but also to predict when assets will reach the end of their life and need replacing. To set the scene for future research, a state of the art review of modeling techniques is presented in Chapter 7.0, which describes techniques that can be used to predict remaining asset life. The techniques are classified into four categories: deterministic models, statistical models, physical probabilistic models, and soft computing or artificial intelligence models. The techniques within each of these categories are described in terms of the theoretical background, practical considerations and examples of application. An attempt is also made at the end of Chapter 7.0 to provide a conceptual map of how water authorities may apply different modeling approaches as their level of asset management sophistication develops.

Research needs are identified at the end of each chapter, and Chapter 8.0 summarizes the most important areas for further research. One of the key recommendations in terms of the WERF research program is the need for a road map that describes the research steps required to facilitate water authorities to identify and then adopt approaches for assessing and predicting asset remaining life that meet their current and future business needs.

More generally, there is an overarching need for industry standardization of terms and definitions, as well as a way to account for obsolescence and redundancy in consideration of asset management and remaining life. Given the importance of risk to both asset management and the concept of remaining asset life, it is important to develop protocols for risk assessment for the main assets types, which consider the technical capacity of water authorities. It also is important to understand data needs for the various techniques and to determine how applicable each technique is at various levels of 'data richness'. A key aspect of this will be to benchmark the different modeling and assessment techniques to determine the extent to which each can reduce the uncertainties in asset management decision-making.

CHAPTER 1.0

INTRODUCTION

1.1 Background

The provision of wastewater services involves numerous assets of many types with design/service lives that range from very short to very long life (from a matter of years, to hundreds of years). The life of many of these assets could potentially be extended if there was a better understanding of what influences asset life; for example, through better knowledge of economic and risk factors, or through improved operational and maintenance practices. Extending the life of assets by even a small proportion has the potential for saving significant amounts of money, and these savings could to be used elsewhere to the benefit of customers and the environment. Management of remaining asset life is thus a key component of the overall asset management challenge.

The need to gain a better understanding of remaining life for clean and potable water sectors in the U.S. was outlined in the U.S. EPA's study of the water infrastructure investment gap (U.S. EPA, 2002a). In this 'gap report', it was noted that better knowledge of the remaining life of capital assets would greatly improve decision-making relating to investment needs for maintaining, upgrading, and expanding infrastructure. A fact sheet produced by the U.S. EPA (2002b) also indicates the central role of remaining asset life (or residual life) to planning the replacement of assets. It lists questions for asset managers to explore when planning asset replacement or rehabilitation, namely:

1. When was the asset installed?
2. What is the expected service life, and where is it in its life cycle?
3. Can the anticipated deterioration rate and eventual failure be predicted?
4. If so, what is the estimated time until rehabilitation or replacement is necessary?
5. Could best management practices and maintenance prevent or extend the time to failure?
6. Can the asset be rehabilitated? How much will rehabilitation cost?
7. If so, would this extend the time to failure? By how much?
8. What will be the incremental life-cycle cost of each alternative?
9. Is the asset technically or commercially obsolete?

Most of these questions relate to the issue of remaining asset life in one way or another, which again underlines the importance of this issue to asset management.

1.2 WERF's Strategic Asset Management Challenge

To better understand remaining asset life and, more broadly, to facilitate asset management practices through targeted research, the Water Environment Research Foundation (WERF) developed a research program referred to as 'Strategic Asset Management Communication and Implementation' (RFPP #06-SAM-1 CO). The general challenge statement for this research program was to:

"Develop Guidance and Decision Support tools for Communicating and Implementing a Strategic Asset Management Program (SAM) for Wastewater (and possibly water) Facilities and Estimating Assets Performance and their Residual Economic Life."

To address this challenge, WERF developed a program of research with the management and project structure illustrated in Figure 1-1. As can be seen, remaining asset life is a key component of the overall SAM research program.

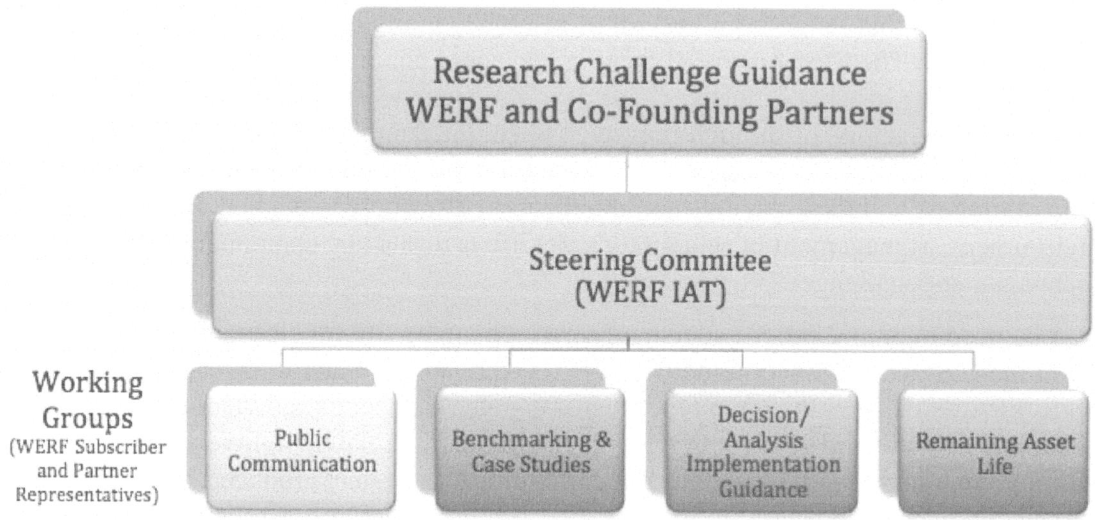

Figure 1-1. The Structure of WERF's SAM Research Challenge.

1.3 Scope of this Report

This report represents initial research undertaken to address the fourth research track (Track 4) of the WERF SAM research program, which addresses 'remaining asset life'. The overall objective of Track 4 is to contribute to the development of techniques, tools and methods for estimating performance and residual economic life of assets. It was proposed that the research be undertaken in a staged manner, so as to provide a stepwise development of concepts and protocols. In summary, the stages of the proposed research were as follows:

1. ***Short term research goals***: including mapping out factors involved with asset deterioration and factors used to give an estimation of remaining asset life, and a state of the art review of the science and practice of "end of life asset management prediction".
2. ***Short to medium term research goals***: including an evaluation of existing methodologies, focusing on key issues including:
   - An investigation into the use of condition-based and performance-based decay curves, including benchmarking of decay curves to real life asset failures, and the examination of the factors upon which the decay curves are based.
   - Determining data needs for the available techniques, addressing data collection protocols, including the application of GIS to parameter data collection.

- Development of protocols for risk assessment for the main asset types (such as pipes) and methodologies for mapping assets onto these protocols using a range of techniques.

3. ***Long term research goals***: including development of more exotic and less commonly used techniques identified during the initial phases of the research, design of an industry database for determining residual lifetimes. This will include the methodology for building upon existing failure models utilizing reliability theory to allow each utility to participate in building up industry wide databases of failure predictions.

This report details the research undertaken to fulfill the short-term research goals. To this end, the research team has produced a synthesis of knowledge in relation to "end of life" and "remaining asset life".

One of the key questions to be addressed in this initial stage of the research was the state of knowledge with respect to the estimation and prediction of remaining asset life, and if there is the capacity to translate between condition and performance data (e.g., the presence of significant defects) and the residual life of an asset. This report builds on previous work by the research team on protocols for condition and performance assessments, as detailed in WERF (2007).

1.4 Project Delivery

Production of this report has involved the following activities:

- A review of the literature and other available information sources;
- A synthesis of knowledge drawing on the project experience of the research team;
- Developing case studies with industry partners to give a practical view of related issues as they apply in different countries (the UK, Australia and the U.S.).

The overall objective of this report is to indicate where further research can be carried out to contribute most effectively to meet the medium to long term objectives of WERF's asset management research program.

1.5 Scope of Assets Covered

The research presented in this report addresses the concept of asset life for all wastewater assets. However, more focus has been given to pipeline (sewer) assets because these represent a large proportion of the value of assets within a wastewater system (it is generally accepted that the value of sewerage is about 80% of the total value of wastewater assets). Sewers are also hidden from view, with a long life and complex deterioration process. As such, estimating remaining life of sewers is a complicated task.

While acknowledging the importance of sewers, this report also addresses non-pipeline assets to some degree. Many of the asset management concepts discussed in the report are generic and equally applicable to pipeline and non-pipeline assets, even though details of the deterioration processes and failure modes differ. As such, while there is less emphasis given to non-pipeline assets, the treatment of these assets is not cursory.

1.5.1 Definition of Asset Adopted

The Infrastructure Management Manual (IPWEA, 2006) defines an asset as "a physical component of a facility that has value, enables services to be provided and has an economic life of greater than 12 months." In practice, the definition of what constitutes an individual 'asset' will vary from company to company, depending on the requirements of their asset management processes. For the purposes of this research, an asset is generally considered to be any piece of equipment that appears on a water authority's asset register/inventory as an individual capital item.

Assets are often complex items, being made up of a number of components. However, the components are not considered to be assets in themselves, unless they are recorded as individual capital items. For example, consider a pump set within a treatment works that is recorded on the asset inventory as an individual capital item, but none of its components (motor, pump, etc.) are listed individually. Within the definition of asset adopted, the treatment work itself would be considered an asset, as can the pump set, but the components of the pump set are not considered to be assets themselves.

1.5.2 Representation of the Asset Stock

Within the wastewater sector, it is natural to consider discrete non-pipeline assets and distributed pipeline assets separately. Discrete non-pipeline assets are generally above ground, contained within a given site, more accessible and easier to assess than pipeline assets. In contrast, pipeline assets are spatially distributed, generally below ground and more difficult to access and assess. Various other descriptors can be used to describe the asset under consideration.

With these issues in mind, and following on from the previous research undertaken into condition and performance assessment strategies and protocols (WERF, 2007), for the purpose of this report, wastewater assets are divided into two main categories:

- Wastewater pipeline assets (rising mains, gravity sewers);
- Wastewater non-pipeline assets (wastewater treatment works, pumping stations and the equipment within these complex assets).

In more detail, the representation of the asset stock used in this project is presented in Table 1-1. As indicated, assets are categorized according to unit type; a unit being defined as the smallest system of components or a section of pipeline considered separately for asset management purposes (the 'unit' is thus the smallest asset in the asset hierarchy). Since there are potentially a large number of unit types within a wastewater treatment system (above and below ground assets), it makes sense to group unit types into common categories. For above ground (non-pipeline) assets, units are categorized as:

- Mechanical and electrical (M&E) assets
- Civil and building (C&B) assets
- Instrumentation, control and automation (ICA) assets

Table 1-1. Representation of the Asset Stock Adopted.

Asset category	Asset	Unit definition*	Other descriptors
Pipeline (below ground)	Laterals (service connections)	Connection	Material, diameter
	Non-critical sewers	Pipe lengths, fitting	Material, diameter, depth
	Critical sewers	Pipe lengths, fitting	Material, diameter, depth
	Man holes	-	Material, diameter, depth
	Valves (block/stop, etc.)	-	Size, type, open/closed
	In-line (underground) storage	-	Size, type
	Force mains	Pipe lengths, fitting	Material, diameter, pumped
	Air valves	-	Size, type
	Surface water outfalls	-	Material, diameter
	Sewer overflow structures	-	Design, size
	Marine outfall	-	Material, diameter, length
Non-pipeline (above ground)	Pumping stations	M&E, C&B, ICA	Capacity
	Detention tanks	M&E, C&B, ICA	Size
	Treatment works	M&E, C&B, ICA	Size, process/complexity
	Stormwater storage	M&E, C&B, ICA	Size
	Sludge treatment works	M&E, C&B, ICA	Size, disposal route
	Other civil structures (roads, walls, etc)	-	Use
	Buildings	-	Use

*A unit is considered to be a sub-system of a larger asset or a section of pipeline considered separately for asset management purposes. Note: **M&E**: Mechanical and Electrical; **C&B**: Civil and Building; **ICA**: Instrumentation Control and Automation.

1.6 Linkage to SIMPLE

In conjunction with other WERF research, the current project will help utilities move towards better practice in asset management. It is anticipated that the research outputs from this project will be integrated into WERF's Sustainable Infrastructure Management Program Learning Environment (SIMPLE) tool, accessible through the WERF website:

http://www.werf.org/am/custom/simple_gateway.cfm (*currently accessible to WERF members only*)

SIMPLE is a web-based knowledge management tool that helps utilities in developing a life-cycle asset management approach. The tool guides utilities on how to determine the most cost-effective investments – acquisition, maintenance, renewal – in their asset portfolio, including how to extend the life of existing assets by implementing optimal maintenance practices and rehabilitation interventions. SIMPLE is based on the process model of asset management illustrated in Figure 1-2, and includes the determination of remaining asset life. (Note: 'remaining asset life' is referred to as 'residual life' in the SIMPLE model of asset management.)

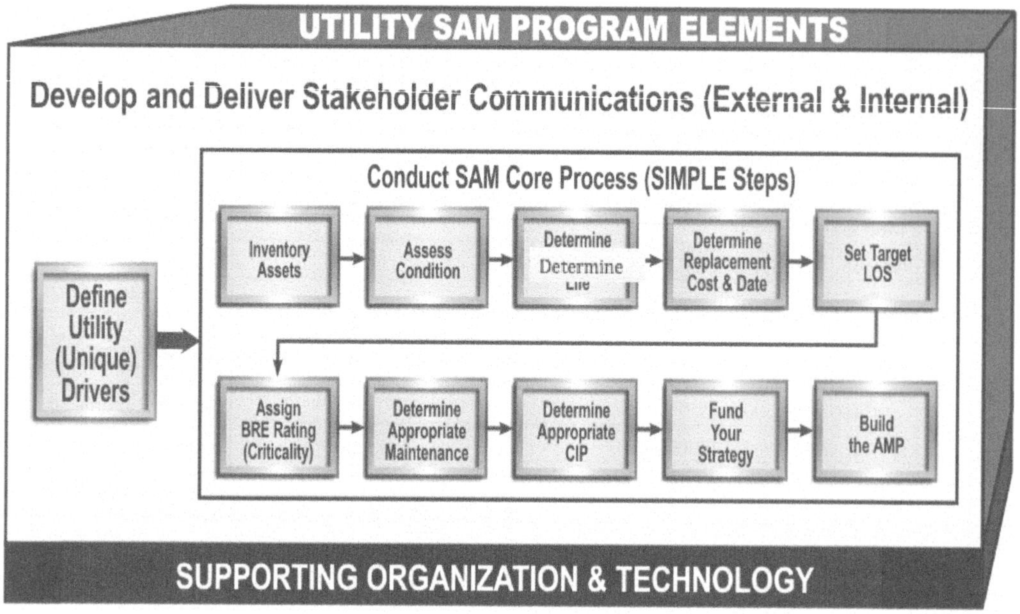

Figure 1-2. The Core Processes of the SIMPLE Asset Management Framework.

1.7 Report Structure

In addition to this introductory chapter, this report is presented in eight chapters:

- Chapter 2.0 provides background information on the concept of remaining asset life.
- Chapter 3.0 presents and discusses factors that influence asset life.
- Chapter 4.0 considers how risks influence remaining asset life.
- Chapter 5.0 provides background information on the role of remaining asset life in asset management.
- Chapter 6.0 discusses approaches to assessing remaining asset life.
- Chapter 7.0 summarizes the state of the art in modeling remaining asset life.
- Chapter 8.0 presents a summary of research challenges associated with the development of pragmatic protocols for estimating and predicting remaining asset life, as well as a suggested research road map.

CHAPTER 2.0

THE CONCEPT OF REMAINING ASSET LIFE

2.1 Summary

General issues relating to asset life are discussed throughout this chapter, with additional details provided in Appendix A. It is shown that asset life and remaining asset life are multifaceted concepts. At a basic level, asset life relates to the repairability of an asset, but is also dependent on a range of other factors. Asset life is modified and extended through maintenance and capital investment.

A taxonomy of asset life and remaining asset life terms is outlined that reflects various viewpoints relating to asset management decisions. Ultimately, however, remaining life reduces to a question of economics; assets should be replaced or retained because it is more economic to do so, so any such taxonomy is by its very nature somewhat arbitrary.

A definition of asset life is proposed that links the end of asset life to the time at which a significant capital (rather than operational) intervention is required. Remaining life is then the time left before a significant capital intervention is required.

2.2 Introduction

This chapter addresses the fundamental question of what remaining asset life actually means in the context of wastewater asset management. While at first this question may seem trivial, on further consideration it becomes clear that the concept of remaining asset life is complex and multifaceted. To illustrate this and to lay the conceptual foundation for later discussions, this chapter addresses the following aspects of remaining asset life:

- Fundamental concepts relating to remaining asset life, including the 'repairability' of assets.
- A taxonomy for asset life and remaining asset life terms that are applicable in the context of asset management.

A summary of anticipated research challenges arising from the information is also presented.

2.3 What is 'Remaining Asset Life'?

The concept of end of life is relatively clear for living organisms; death occurs when an organism's biological functions cease and we define end of life in these terms. The situation is less well defined when the end of an asset's end of life is considered. For example, many assets are repairable, so loss of function does not imply end of asset life. At the other extreme, an asset can be considered at the end of its life well before any failures actually occur, due to the relative balance between risk and replacement or rehabilitation costs.

The issue of end of asset life is further complicated when assets include a number of components. As discussed in Chapter 1.0, an asset is generally a complex item made up of a number of components, each of which can have very different expected lives. Some parts of an asset will thus commonly reach the end of their life before others, and replacement of these components will allow the remainder of the system to continue in service. Such replacements form part of the necessary and expected maintenance for such an asset.

The term maintenance covers a broad range of planned or unplanned activities for preserving asset condition (Vanier, 2001). Generally speaking, maintenance tasks (such as repair, re-alignment, and oil change) are intended to slow the overall deterioration of an asset and extend its life. The impact of deterioration and its relationship to end of asset life is considered in more detail throughout this report (e.g., see Chapter 3.0). For the purposes of the current discussion, however, it is worth considering the overall process of deterioration, how this is addressed through maintenance activities and what this means to asset life.

To this end, Figure 2-1 shows schematically the deterioration in asset condition over time, when the deterioration is regularly addressed by maintenance. As shown, periodic maintenance has the effect of restoring condition, but the level of condition achieved is often below that of a new asset. More extensive refurbishment and upgrading are undertaken during the life of many wastewater assets to replace components and to change the asset's functionality and accommodate changed service requirements. Through periodic maintenance and capital expenditure made throughout its lifecycle, the life of an asset is thus extended and modified (NSW, 2001). Ultimately, end of asset life occurs when the asset is replaced. However, given the maintenance options available, predicting when this event will occur can be a complex issue.

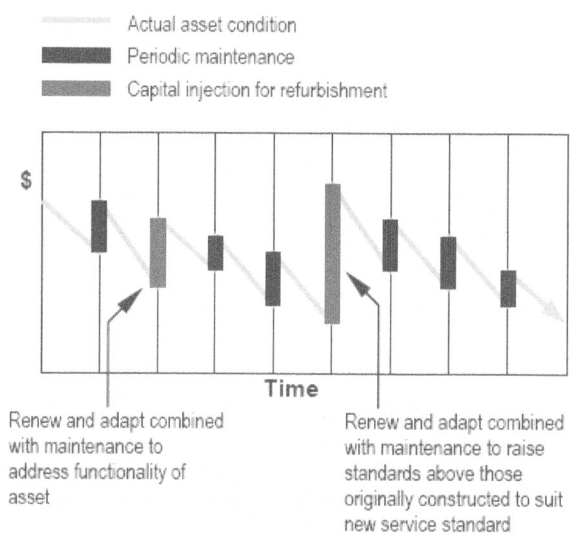

Figure 2-1. The Impact of Maintenance on Asset Life.
(Reprinted with permission from NSW, 2001)

Figure 2-1 indicates that an asset may be subject to periodic maintenance. For many types of assets, this periodic maintenance will involve planned activities intended to prevent or mitigate asset failure (e.g., Arunraj and Maiti, 2007). Maintenance can also involve repair of failed components, although economical repair options are not available for some assets. Whether an asset can be repaired or not thus has a significant influence on the concept of remaining asset life.

2.4 The Issue of Repairability

From the perspective of reliability theory, a non-repairable asset or component is one that is discarded and/or replaced after its first failure. A repairable asset or component is one that, when a failure occurs, can be restored to an operating condition by some repair process (e.g., Basu and Rigdon, 2000; Watson et al., 2001). In practice, however, economic and other considerations blur the distinction between the two types of assets.

2.4.1 Non-Repairable Components

A light bulb is an obvious example of a non-repairable component. On failure, a light bulb must be replaced since there is no economic repair option available. At a fundamental level, the concept of remaining asset life can be relatively straightforward for non-repairable assets or components; when the asset or component fails, it is at the end of its life.

For some assets, however, it may not be acceptable for the asset/component to fail due to the consequences that would be incurred, and other approaches to defining end of life must then be used. For example, condition/performance monitoring approaches can be used when the on-set of failure can be detected in some way (see WERF, 2007 for more details). Alternatively, when the on-set of failure cannot be detected, components can be replaced early using some measure of age or usage (for example, a car engine timing belt is replaced according to its age or number of miles driven).

2.4.2 Repairable Assets

Many of the complex assets operated by water authorities are repairable, even though individual components may not be. For repairable assets, end of life is not explicitly related to failure, since function can be restored through some kind of repair. Whether or not repair is more economical than replacement is a separate consideration. Various economic factors must be considered at the time of failure to determine if the asset should be repaired or replaced. These factors can be considered within a risk-informed framework, as discussed in Chapter 3.0.

As with non-repairable components, it may be undesirable for a repairable asset to fail due to the consequences of the failure. As noted previously, approaches such as condition and performance monitoring provide a means of assessing when an asset or component is reaching the end of its life. Early replacement also can be undertaken where the on-set of failure cannot be detected in this way.

2.4.3 Overlap Between the Definitions

In practice, the issue of repairability depends upon the definition of assets adopted and the context within which repairs are considered. As noted above, assets involved in the supply of wastewater services are generally complex and consist of various components. When failure of an individual component occurs, such that the asset itself fails in someway (e.g., partial or full loss of some or all functions), it is generally possible to replace the component to restore functionality. In such cases the asset would be considered repairable, and the component non-repairable.

Economic considerations can, however, effectively blur the distinction between repairable and non repairable assets. For example, it might be technically possible to repair a small (and thus cheap) pump but, due to the cost of repair being greater than the cost of purchase, it would probably be discarded and replaced (WERF, 2007). Similarly, Elliot et al. (AWWARF,

2001) noted that prior to initiating test procedures on electric motors, it is necessary to compare the cost of replacing the motor to the cost of the testing. The authors note that for some small-commodity-size (less than 25Hp) motors, it is cheaper to replace than to completely evaluate and repair. Motors of 25 Hp and larger may or may not be cheaper to replace outright instead of evaluating and repairing. Assets that can be replaced at a cost less than that of the repair can be considered non-repairable for all intents and purposes.

2.5 A Taxonomy of Remaining Asset Life

As discussed in the previous section, remaining asset life is not a simple concept. One implication of this is that the term 'remaining asset life' is imprecise and will mean different things in different contexts. For example, Vanier (2001) discusses the concept of asset life, and definitions are presented for various related terms. In particular, ***service life*** is defined as "the actual period of time during which [the asset] or any of its components performs without unforeseen costs of disruption for maintenance and repair". The term "unforeseen" is a key word in this definition, since many assets must be maintained to ensure that the service life is reached, as illustrated previously in Figure 2-1. Vanier (2001) also notes that assets or components are sometimes replaced at the end of their ***technical service life*** (i.e., when they no longer meet functional requirements), but can also be replaced when they reach their ***economic service life*** (that is, when it is less expensive to replace than to maintain the asset).

From this brief discussion, it can be seen that a range of definitions are relevant to any discussion of asset life. In an attempt to rationalize terminology, a review of asset life terms was undertaken, building on the experience of the research team. The following taxonomy of terms provides insights into different views of asset life:

- ***Design life:*** the period of time over which the asset is designed to be available for use and able to provide the required level of service at an acceptable cost and level of risk. As noted in the International Infrastructure Management Manual (IPWEA, 2006), the anticipated life of an asset or component can be measured in terms of time, number of cycles, distance intervals, etc. Lillie et al. (2004) refer to the concept of 'design service life' (the authors focus on pipes, but the concepts are more generally applicable), as the life that was expected when the asset and system were designed, considering expected loading regimes, environmental protection and deterioration. In practice, many assets can operate well past the end of design life due to fundamental redundancies built in at the design stage.

- ***Service life:*** the period of time over which the asset is actually available for use and able to provide the required level of service at an acceptable risk; e.g., without unforeseen costs of disruption for maintenance and repair.

- ***Operational life:*** some assets may be operated past the point where they provide required level of service at an acceptable risk (i.e., operated past the asset's service life). As such, the operational life is taken to be the time over which the asset remains operational irrespective of performance, risk or cost considerations.

Similarly, various definitions relevant to the ***end of asset life*** can be compiled, and the following taxonomy of terms is considered to provide useful insights:

- ***End of physical asset life:*** when the asset is physically derelict and non-functioning; for example, a sewer that has collapsed to the extent that function is no longer being provided and repair is not an option.
- ***End of technical service life:*** when the asset is failing to provide required functionality, service levels and/or reliability.
- ***End of economic asset life:*** when the asset is physically able to provide a service, but ceases to be the lowest cost alternative to satisfy a particular level of service (IPWEA, 2006). In practice, this often reduces the time when the risk/cost associated with retaining an asset exceeds the cost of rehabilitating the asset (risk issues are considered in detail in Chapter 4.0).
- ***End of financial life:*** when the asset's initial capital value is fully depreciated.
- ***Obsolescence:*** when the asset is obsolete because of changes in technology, regulatory requirements, or performance criteria.

Remaining life is simply the time remaining until one of these end of life criteria are reached. The different perspectives of asset life and remaining asset life are described more fully in Appendix A.

2.5.1 A Holistic Definition of Remaining Asset Life

While various definitions of remaining asset life can be identified, for the purposes of on-going research it is desirable to have a single holistic definition. Regardless of the definition used for end of asset life, the outcome is that there is a significant level of investment made in the asset; either to rehabilitate it or to replace it.

As indicated in Figure 2-1, renovation of the asset will extend the original asset life. However, after renovation the asset will have a new design life that commences from the completion of the intervention. It thus is useful to define end of asset life in terms of when an asset will need a significant (capital, rather than operational) investment. As such, the researchers define end of asset life and remaining life in the following way:

> *End of asset life is the time at which a significant (capital rather than operational) investment is made. Remaining life is then the time left before a significant capital intervention is required.*

This definition is somewhat similar to the concept of 'phase of service life' used by Li et al. (2007), referring to a time period within the whole service life at the end of which maintenance such as repairs, strengthening, or rehabilitation are required. The definition of end of life adopted herein differs from this concept only in that the asset must be undergoing a significant capital (rather than operational) intervention before it is deemed to be at the end of its life. This definition aligns with the needs of asset managers, who must determine when an asset is reaching the end of its life because investment is then necessary and must be catered for (unless the asset is no longer needed, in which case the lack of on-going need drives the consideration of end of asset life).

2.6 Research Challenges: Asset Life Concepts

From the information presented in this chapter, the challenge associated with assessing remaining life can be summarized thus:

1. The taxonomy and terminology used herein are only indicative. Industry standardization of terms and definitions would be of benefit to the sector.
2. There is also a need to determine how to account for obsolescence in consideration of asset management planning and remaining life assessments (for example, changes in design approaches and modification to assumptions made in models).

CHAPTER 3.0

FACTORS THAT INFLUENCE ASSET LIFE

3.1 Summary

Factors that influence the life of various asset classes are presented and discussed in this chapter. Those factors relating to physical life are considered in asset life cycle terms, since decisions made in each part of the asset life cycle can influence the useful life obtained. Various other factors are not related to the asset themselves, but are to do with the economic aspects of the water authority and/or level of service the authority has to deliver to customers, the community and the environment.

Specific factors relating to the lives of various generic asset types are tabulated and discussed for generic asset categories: i.e., sewer; civil and building; mechanical and electrical; and instrumentation, control and automation assets.

For each asset category, factors are identified that relate to design and construction, operating environment, operation and maintenance, external factors and economic and service factors.

A summary of the characteristic features of the end of asset life for each asset category is also presented, along with a review of key performance indicators (KPIs) that are used to monitor asset service/function.

3.2 Introduction

As indicated in Chapter 1.0, one of the aims of this initial stage of the research was to map out factors that influence remaining asset life, including factors that influence deterioration of various types of asset. This chapter presents information on:

- Generic groups of factors that influence the deterioration and remaining life of all asset types.
- Specific factors relating to the lives of various generic asset types.
- Anticipated research challenges.

The factors presented have drawn upon a review of the available literature, including project reports, and the experience of the research team.

3.3 Generic Categories of Factors that Influence Asset Life

From a purely physical (rather than economic) perspective, asset life is influenced by decisions made at various stages of the asset's life cycle, and generic factors that influence an asset's physical life can thus be categorized in life cycle terms; for example:

- ***Design and construction/installation factors***: the quality of the original asset, including design criteria and standards, materials of construction, quality of construction/manufacture, installation and commissioning, etc.

- ***Operating environment factors***: the environment in which the asset is constructed and subsequently operated; including the quality of wastewater conveyed or retained, and the operational demands (loads, frequency of use, etc.) placed on the asset.
- ***Operation & maintenance factors***: the quantity and quality of operation and maintenance (especially in relation to manufacturer's recommendations and industry best practice).
- ***External factors***: including vandalism, third party damage, weather and ground conditions.

From an economic and service perspective, a range of other factors must be considered, many of which will be external to the asset, its design, construction and operation. Categories of external factors that influence an asset's economic or service life include:

- ***Required performance standards***: including reliability and other service/performance metrics that may change over time.
- ***The consequences of failure***: costs, operational, service provision, environmental impact, social impact, etc.
- ***Cost of repair, replacement or rehabilitation***: in particular, changes to the cost of repair, replacement or rehabilitation will impact the timing of when an asset is considered at the end of its life.
- ***Available budgets***: ultimately, the level of funding a water authority receives can have a significant impact on when assets are considered at the end of their life. All other things being equal, a water authority may wish to replace an asset, but if there are insufficient budgets, it may be forced to carry the cost of operating aged assets. In many instances, this type of consideration is intimately linked with the ability to convince regulators, customers and other stakeholders of the need to renew assets.
- ***The rate of technical innovation***, which in turn impacts the rate of technical and functional obsolescence.

All the above factors, whether related to the asset's physical or economic life, can be considered within a holistic risk framework, as discussed further in Chapter 3.0. With such a risk framework in mind, two other generic factors can be considered to influence the economic life of an asset:

- ***The failure mode***: e.g., assets with sudden/catastrophic failure modes will be considered differently to assets that have gradual, non-catastrophic failure modes.
- ***Cost of failure mitigation***: i.e., operational risk management, including contingency planning and emergency response.

3.4 Service Lives of Wastewater Assets

The generic categories of factors given above apply to all these asset types to one degree or another, but the specifics and relative importance of factors are different for each asset type, and also differ according to the operational context of individual assets. Nonetheless, wastewater assets are considered herein in terms of generic asset categories so that the broad range of asset types used in the provision of wastewater services can be covered in one report. As noted previously, the generic asset categories considered are:

- Sewerage assets (gravity sewers and forced mains)
- Civil & building (C&B) assets
- Mechanical & electrical (M&E) assets, and
- Instrumentation, control and automation (ICA) assets (which are taken to include telemetry assets).

Specific factors relating to the lives of various generic asset types are tabulated and discussed in the following sections for these generic asset categories, presented according to the categorization scheme given in the previous section (i.e., factors are identified that relate to design and construction, operating environment, operation and maintenance, external factors and economic and service factors).

A summary of the characteristic features of the end of asset life for each asset category is also presented, along with a review of key performance indicators (KPIs) that are used to monitor asset service/function.

3.5 Sewerage Assets

Sewerage assets are categorized in a number of ways; for example, by mode of conveyance (i.e., gravity sewers and pumping/forced mains), as well as by size and function (i.e., reticulation, branch, and trunk sewers of various diameters).

Gravity sewers provide drainage of wastewater and/or stormwater; modern systems being generally separate, with some older systems carrying combined waste and stormwater. Reticulation gravity sewers convey local sewage discharges under gravity flow to trunk sewers and/or pumping stations. Trunk sewers carry bulk flows to larger interceptor sewers or to treatment works.

The only failure mode that represents the end of 'physical life' of a sewer is structural collapse; i.e., when an asset is so structurally deteriorated that it loses arching support and can only be replaced or rehabilitated with a structural method. All other failure modes can mean the asset is at the end of its life, but this determination will depend on service and economic considerations.

Pumped mains (also known as forced or rising mains) convey sewage under pressure from a pumping station to a point of discharge, such as a gravity sewer or a sewage treatment works. As pressure pipes, forced mains tend to fail structurally, leading to bursts or long term leaks. These pipes are commonly constructed from plastic and ferrous materials. In the past, some rising mains were also constructed from asbestos cement.

Gravity sewers and pumping/forced mains are generally long-lived assets, but characteristic lives depend on material of construction.

Characteristic features for end of asset life for gravity sewers and forced mains are summarized in Table 3-1. Davies et al. (2001a) list a range of factors that influence the structural deterioration and collapse of rigid sewers, and these have been extended to include factors relevant to other pipe materials (Table 3-2) and to the case of forced mains (Table 3-3).

Table 3-1. Characteristic Features of End of Life for Sewerage Assets.

Asset Category	*Characteristic Features of End of Life*
Gravity Sewers	Lack of structural integrity (collapse, infiltration, exfiltation) Deformation (loss of arching support) Loss of level Capacity issues (future) Service issues (including blockage rate, flooding, etc.)
Forced Mains	Lack of structural integrity (burst, leakage) Capacity issues (future)

Table 3-2. Factors that Influence the Life of Gravity Mains.

Category	**Factors**
Design and Construction	Manufacturing quality (pipes, joints, etc.) Sewer pipe material (and appropriateness for operating context) Joint type and material Pipe section length Connections (design details, number, etc.) Sewer size, depth and shape Structural design Asset protection (lining, etc.) Installation method Standard of workmanship Bedding material and type Soil/backfill type
Operating Environment	Surface use Surface type Surface loading (including construction traffic) Traffic characteristics Ground water level Infiltration/Exfiltration Sewage characteristics Sediment load Surcharge potential
Operation and Maintenance	Inappropriate maintenance Maintenance approach adopted ('do nothing' through preventive maintenance) Damage to the pipe when a subsequent connection was made
External Factors	Water main bursts/leakage Ground movement (subsidence) Maintenance of other buried assets Root interference Climatic and related soil interactions (shrink/swell) Illegal connections Third party interference/damage Changes in population served or water use (i.e. changes in flows)
Economic/Service Factors	Cost of repair, rehabilitation or other options Technical feasibility for trenchless technology (especially structural state of host sewer) Stakeholder requirements with respect to service Potential/actual Public Relation impacts (i.e. level of community outrage) Health and environmental concerns Customer complaints Available budgets (may be less than for water assets) Level of service and risk measures

Table 3-3. Factors that Influence the Asset Life of Pressure Mains.

Category	Factors
Design and Construction	Manufacturing quality (pipes, joints, etc.) Pipe size and pressure class Pipe depth Pipe material Pipe section length Joint type and material Design of discharge points (e.g., evolution of gases) Asset protection (lining, cathodic, etc.) Installation method Standard of workmanship Bedding material and type Soil/backfill type
Operating Environment	Surface use Surface type Surface loading (including construction traffic) Traffic characteristics Sewage characteristics (BOD/COD, temperature, fat content, etc.) Sediment load Flow conditions (scouring velocities, stagnation, etc.) Detention time Operating and surge pressures Areas of changing pressure (e.g., discharge points subject to H_2S release/attack)
Operation and Maintenance	Valve operation (e.g., rate of closing) Pump operation Inappropriate maintenance Maintenance approach adopted ('do nothing' through preventive maintenance)
External Factors	Water main bursts/leakage Ground movement Maintenance of other infrastructure assets Root interference Climatic and related soil interactions Third party interference/damage Stray electrical interference currents
Economic/Service Factors	Cost of repair, rehabilitation or other options Stakeholder requirements with respect to service Potential/actual Public Relation impacts (i.e., level of community outrage) Health and environmental concerns/impacts Available budgets Level of service and risk measures

3.5.1 Factors that Influence the Physical Life of Sewers

Design and Construction

Design and installation issues are critical for all pipeline assets. Design decisions relating to pipe size, material, joint type, type of bedding and backfill, pipe route/emplacement, etc. can all have significant impacts on the life of an asset.

Forced mains must resist static, dynamic and transient (surge) pressures. Design of the main and pumping system must be appropriately matched. For example, some early uses of plastics materials led to failures of pressure mains due to surge and cyclic pressures.

Operating Environment

Many materials used in the construction and jointing of sewers deteriorate over time through processes such as corrosion or mechanical wear. This type of deterioration begins immediately after commissioning due to the harsh internal environment of carried sewage and, sometimes, the aggressive external soil environment (Zhao, 1998). Deterioration can also occur following initial damage to the asset, which is worsened by external loading and operating conditions.

The load-bearing capacity of a gravity sewer is related to both the structural strength of the pipe itself, and the strengthening effect imparted by the surrounding soil (Davies et al., 2001a). Loss of soil support associated with poor compaction or due to infiltration will exacerbate the progression towards failure. For gravity sewers, there are thus two basic classes of failure mechanisms:

- ***Degradation of the materials*** from which the sewer is constructed, either of the pipe body (due to such processes as carbonation of cementitious materials, chemical attack due to acidity, or septicity and corrosion of ferrous materials) or of joints (through such processes as settlement cracking of cement mortar joints, erosion of clay puddle joints, or degradation of elastomeric joint rings).
- ***Deterioration of the structure*** caused by loss of soil support around the pipe due to migration of soil resulting from a construction or manufacturing defect.

In general, the overall process of deterioration is characterized as one of initiation-progression-propagation-failure (e.g., see Davies et al., 2001 and references therein):

- ***Initiation***: an initial, often minor, defect is imparted in some way.
- ***Propagation***: deterioration then occurs through various mechanisms, such as the loss of soil support from around the sewer, which reduce the structural capacity of the sewer.
- ***Failure***: when it eventually occurs, failure is often triggered by some random event that may not be related to the deterioration process.

The length of time taken to progress through these stages depends on the asset material and the characteristics of the surrounding soil; for example, brick sewers in fine (running) sand can deteriorate very quickly once a crack is large enough to allow supporting soil to be washed into the sewer. The level of ground water will also have an effect.

Forced mains also are subject to degradation of material and other deterioration processes. In addition to the external loads imparted on gravity mains, forced mains must also resist static, dynamic and transient (surge) pressures. As a pressure main deteriorates, the probability of failure increases as the main becomes more susceptible to the effects of events such as pressure surge. Final failure is again likely to be the result of a combination of deterioration in condition and the influence of the internal and external loads; i.e. failure, is often triggered by a particular event or combination of events.

Operation and Maintenance

Closed-circuit television (CCTV) inspection of gravity sewers allows remaining life to be estimated from the presence of critical defects like deformation, cracks and fissures. Early repair of these defects using non-structural lining and patch repair techniques can be an economic means of extending an asset's life. Other maintenance activities—like sediment removal, jetting, and root cutting—are related to service issues, rather than physical life.

As noted above, rising mains must resist static, dynamic and transient (surge) pressures. Operation of pumps and valves so as to avoid peak surge pressures can thus greatly influence asset life, especially for pipe materials such as PVC that are vulnerable to cyclic pressure.

External Factors

Being spatially distributed, pressure mains and gravity sewers are vulnerable to a range of external influences, including third party damage associated with digging activities, and the impact of nearby infrastructure (maintenance, loading and leaks/bursts). The presence of trees and soil expansion/contraction also influence the rate of deterioration.

Changes in static and dynamic loads (e.g., from construction work, buildings and road traffic) also can accelerate deterioration of sewer assets. In addition, peaks in dynamic loads can trigger ultimate failure.

3.5.2 Factors that Influence the Economic/Service Life of Sewers

General Comments

As the physical structure of a sewer deteriorates, it will eventually reach the point where service or economic considerations drive rehabilitation or replacement. Other factors not specifically related to the physical deterioration process are also important, and these tend to be related to asset location and/or operational/functional context.

A key service issue for many smaller diameter mains (e.g., < 300mm/12") is the rate of sewer blockage. Blockages can occur due to a range of factors including fats, oils and grease build up; accumulation of silt; introduction of 'foreign' objects (debris, nappies, etc.); and root intrusion. Blockages by root intrusion can be viewed as a combined consequence of poor water tightness (i.e., there must be cracks in the pipe body or leaking joints for roots to get in) and vegetation in the area.

Another key service driver for replacement of sewers is the frequency and type of flooding events related to sewer capacity, especially flooding of properties and discharges that impact watercourses. However, all types of flooding represent a health and environmental risk and can thus drive replacement of an asset. Often a consideration is the level of public relations impact and/or the amount of outrage generated by an event. In particular, the impacts generated by the failure of forced mains can be high because the sewage is conveyed under pressure.

Service/Performance Measures

Poor performance of a sewer, group of sewers or within a management area (e.g., drainage areas) measured through various Key Performance Indicators (KPIs) will influence the decisions to replace or rehabilitate assets, and thus the issue of remaining asset life. The following Key Performance Indicators (KPIs) are relevant to the remaining life of sewers:

- Level of Inflow/Infiltration/Exfiltration
- Sewer blockages, especially repeat blockages
- Dry weather flooding due to insufficient capacity (internal and external to properties), especially repeat floods
- Wet weather flooding (internal and external to properties), especially repeat floods

- Properties affected by flooding incidents, and properties at risk of flooding
- Flooding events due to asset failure (i.e., sewer collapse)
- Sewer overflow discharge (frequency and volume), especially discharges to watercourses (including over-land floods that impact watercourses)
- Pollution incidents at sewer overflow structures
- Number of sewer collapses
- Customer complaints (related to odor, blockages or flooding events)

3.6 Civil and Building Assets

For the purposes of this report, civil assets are taken to be civil structures within treatment works and pumping stations, such as containment vessels, tanks or channels (sewers or rising mains are not considered to be civil structures). Materials of construction have traditionally been bulk or reinforced concrete, with some brick structures, though there is also some use of prefabricated structures, both for process vessels and buildings. These prefabricated assets are made from a range of materials including steel and plastics, including polypropylene, polyethylene (PE) and glass reinforced plastic (GRP).

Buildings used to house plant or site personnel are often constructed of brick or concrete, but can also be prefabricated.

Characteristic features for end of asset life for civil and building assets are summarized in Table 3-4, and factors that influence asset life are summarized in Table 3-5 and discussed in the sections that follow.

Table 3-4. Characteristic Features of End of Life for C&B Assets.

Asset Category	*Characteristic Features of End of Life*
Civil	Unsound structure (structure considered unsafe) Excessive maintenance required Capacity issues
Building	Lack of structural integrity (building considered unsafe) Excessive maintenance required Building not adequate in terms of size, location and/or facilities

3.6.1 Factors that Influence the Physical Life of C&B Assets

Design and Construction

Provided the structures are correctly designed, built in accordance with the design and well maintained, the service life of C&B assets will generally meet or exceed the design life.

Treatment works and pumping station assets are not normally allowed to fail. Process assets such as tanks and digesters are thus often installed in a configuration so that performance can be maintained if an item of equipment fails, or when an asset or process stream needs to be taken off-line for maintenance. The level of redundancy built in at the design stage can have a significant impact on the level of risk than can be accepted, and thus on the concept of remaining asset life.

The expected lifetime of prefabricated structures is usually much shorter than for the equivalent structure constructed of brick or concrete. Additionally, the rate of deterioration and potential for premature failure is significantly influenced by the level of care exercised during transport and assembly of the prefabricated sections. This is particularly true for tanks made from coated steel plates.

Table 3-5. Factors that Influence the Life of C&B Assets.

Category	Factors
Design and Installation	Structural design Materials of construction or manufacturing quality, as appropriate Detail design features (joints, weather proofing, corrosion protection, etc.) Asset protection (lining, cathodic, etc.) Installation method Standard of workmanship
Operating Environment	Sewage characteristics (including septicity) Structural loads (including hydrostatic and hydrodynamic loads) Exposure to aggressive elements such as pollution, salt spray, etc.
Operation and Maintenance	Inappropriate or insufficient maintenance Maintenance approach adopted ('do nothing' through preventive maintenance)
External Factors	Ground movement (subsidence) Weathering Third party interference/damage (including vandalism) Security requirements
Economic/Service Factors	Level of asset redundancy Cost of repair, rehabilitation or other options Available budgets Level of service and risk measures Capacity issues Obsolescence due to changes in process design

Operating Environment

The failure mechanism of civil assets and buildings is generally through the gradual loss of mechanical strength due to deterioration processes. The exception to this is where there is sudden catastrophic failure due to outside factors such as subsidence or seismic activity. Ultimately, failure occurs when the structure is no longer able to support the loads imposed on it, but earlier 'trigger' points will usually be considered to indicate the end of asset life. For example, a trigger point could be when spalling occurs or there is leakage from a structure.

Deterioration of C&B assets can occur through degradation of materials of construction; for example, carbonation due to carbon dioxide penetrating the concrete, chloride attack due to chloride salts from sea spray or de-icing salts, and corrosion of steel components and reinforcement. Steel tanks are susceptible to corrosion of the panels and failure of the seals between the panels.

In a wider context and similar to sewers, the deterioration of civil and building structures can often be described in three stages: initiation (an initial, often minor, defect is imparted in some way); propagation (deterioration through various mechanisms), and failure. As with sewers, failures often are triggered by some random event that may not be related to the deterioration process.

During the initiation stage, the likelihood of structural failure remains essentially the same as when the structure was built. During the propagation stage, the likelihood of failure will increase, but in most structures there is a degree of inherent "structural redundancy". For example, reinforced concrete structures have more reinforcing steel to control cracking of the concrete during curing than is required to meet the structural loads. In addition, the design will include safety factors. Hence, the likelihood of failure remains low during the propagation stage until any "structural redundancy" has been used up. After this time, the likelihood of failure rapidly increases as further deterioration occurs and may progress so far that rehabilitation of the structure is no longer economically feasible.

Operation and Maintenance

Civil and building assets require infrequent maintenance, and are generally replaced for functional or operational/capacity reasons; e.g., when they are no longer required or are too small. They are rarely replaced because of structural failure but extensive renovation may be undertaken during their service life.

Building assets require on-going maintenance to prevent structures from deteriorating. Aesthetic, health and safety, and security considerations may also require extensive renovation of building assets at some stage in their service life.

External Factors

Civil and building assets are impacted by external factors such as weather, local environment (e.g., sea spray), ground movement (due to subsidence or seismic activity), and third party interference, especially vandalism. In particular, steel tanks are vulnerable to third party damage including damage to corrosion prevention coatings.

Security considerations (for buildings and other structures intended to keep out intruders) can have an impact on end of asset life considerations.

3.6.2 Factors that Influence the Economic/Service Life of C&B Assets

General Comments

Civil assets generally have long lifetimes and rarely fail structurally. They are thus often replaced due to inadequate capacity, or as a result of modifications to the treatment process required to meet changes in effluent standards, rather than due to deterioration or failure of the structures.

Building assets also rarely fail structurally, and deterioration is often an issue of aesthetic appearance rather than functionality. However, assets may become functionally obsolete due to changes in demands or staffing levels. Hence, buildings can be considered at the end of their life when the design is not adequate for its current or intended use in terms of size, location and facilities.

Service/Performance Measures

For the majority of civil assets there are no specific KPIs that can be used to measure performance of the civil asset itself. This is because deterioration of the asset tends not to have an impact on the process or function until the structure has failed (i.e., it has become unsafe or collapsed). Condition assessments can be used to track the progression of deterioration through inspection of defects such as cracks, spalling, corrosion of reinforcement bars, depth of cover to reinforcement, depth of carbonation attack and depth of chloride attack.

KPIs can, however, be used to measure the performance of a civil asset in terms of capacity requirements; for example:

- Hydraulic adequacy and/or process capacity at all flows
- Headroom with respect to inefficiencies in upstream/downstream processes

- ♦ Distribution of flows between and within assets
- ♦ Process retention times

Building assets are required to provide functions that can be measured through KPIs; for example:

- ♦ Energy efficiency
- ♦ Overall usage/occupancy
- ♦ Measures of functionality (water tightness, security, etc.)

Poor performance of a C&B asset or group of assets measured through these KPIs will influence the decision to replace or rehabilitate assets, and thus the issue of remaining asset life.

3.7 Mechanical and Electrical Assets

Mechanical and Electrical (M&E) assets are the hardware within pumping stations, treatment works, etc. that underpin many of the processes and functions. Examples of these types of assets include pumps, motors, valves, compressors, interprocess pipework, tank bridges and scrapers, automated control valves, mixers, and electrical power generators.

Characteristic features for end of asset life for M&E assets are summarized in Table 3-6, and factors that influence asset life are summarized in Table 3-7 and discussed in the sections that follow.

Table 3-6. Characteristic Features of End of Life for M&E Assets.

Asset Category	Characteristic Features of End of Life
Mechanical	Excessive number/frequency of failures/breakdowns. Excessive maintenance costs being incurred. Health and safety issues. Obsolete Inoperable
Electrical	Excessive number/frequency of failures/breakdowns. Excessive maintenance costs being incurred. Electrically unsafe. Obsolete Inoperable

Table 3-7. Factors that Influence the Asset Lives of M&E Assets.

Category	Factors
Design and Installation	Manufacturing quality Inappropriate design of components (specification or design details) Inappropriate design of process (e.g., inadequate grit removal and screening) Asset protection (ICA, environmental protection standards relating to dust and moisture, etc.) Installation method Standard of workmanship Misalignment and vibration (shafts, bearings, etc.)

Category	Factors
Operating Environment	Exposure to elements Ambient temperature Humidity Sewage characteristics (including strength, presence of foam, fats or salts) Sludge characteristics Water damage Operating temperature Cavitation of pumps Leakage Over-current or over-voltage events (i.e., higher current/voltage compared to rating) Power supply (variations, interruptions, etc.) Detention time (leading to septicity) General housekeeping (items free of dust, dirt, etc., as far as is practicable)
Operation and Maintenance	Misalignment and vibration (shafts, bearings, etc.) Incorrect maintenance procedures (oiling, etc.) Level of preventive maintenance Slack on drive belts Operational factors (start up procedures for pumps, operating speed, valve operation) Control relays chatter Electrical arcing Overheating
External Factors	External damage Human error (e.g., cumulative impact of incorrect operation or maintenance procedures)
Economic/Service Factors	Cost of repair, rehabilitation or other options Potential/actual Public Relation impacts (i.e., level of community outrage) Health and environmental concerns Customer complaints Available budgets Regulatory changes (levels of service required) Changes to processes or operational requirements (including capacity issues) Obsolescence (due to more energy efficient assets, spares availability, etc.) Level of asset redundancy

3.7.1 Factors that Influence the Physical Life of M&E Assets

Design and Installation

Mechanical and Electrical assets are generally made up of a range of components. Individual components may not be repairable, but overall the M&E units generally are repairable. Sudden early life failures may occur where there are “hidden” manufacturing defects or a component has been incorrectly designed or specified. Detailed design of the assets at the component level thus impacts the expected life of the overall unit.

At the process level, process streams are not normally allowed to fail. Critical assets such as pumps and compressors are thus often installed in a ‘duty’ and ‘stand-by’ configuration so that performance can be maintained if an item of equipment fails or needs to be taken off-line for maintenance.

The level of redundancy built in at the design stage can have a significant impact on the level of risk than can be accepted, and thus on the concept of remaining asset life.

Operating Environment

M&E assets can be exposed to aggressive operating environments, or may be housed in relatively clean areas.

For mechanical assets, end of physical life is often caused by gradual wear of the moving components or mechanical failure. The rate of deterioration can be affected by factors such as:

- The quality of the design and manufacture of the equipment;
- The way in which the equipment is serviced;
- Operating conditions including changes in usage, and additional loading, etc.
- The operating environment (e.g., sewage pumps have a much shorter life than water pumps).

The level of wear is dependent on both the operating environment and the frequency/duration of use. An increase in the mechanical load can also increase the rate of wear on moving parts and this can have a disproportionate impact on the remaining life.

Electrical assets are subject to deterioration due to environment (moisture and ambient temperature) and self-generated heat, which can degrade materials such as insulation. Deterioration/corrosion of electrical components (switches, connectors, etc.) also occurs due to ingress of moisture.

Operation and Maintenance

M&E assets are generally repaired or replaced because either their performance has declined to an unacceptable level or they have ceased to function. Periodic maintenance of components subject to wear will maintain the operational efficiency and prolong operational life.

Correct maintenance and operation of assets can extend the lives of M&E assets significantly, especially in comparison to the expected life if operation and maintenance are incorrectly carried out. In particular, carrying out routine maintenance activities such as greasing and lubricating moving parts will prolong asset life. Undertaking preventive maintenance likewise can be expected to have a significant positive impact on asset life.

Any change in an asset's operational context has the potential for changing the remaining life; i.e., changes to maintenance or operational procedures can increase or decrease remaining life, depending on what the change is. For example, increasing the duty cycle of a motor or pump leads to an increase in deterioration. In contrast, the level of monitoring/inspection and the ability to diagnose problems (interpreting the results of the monitoring effort) before they result in catastrophic failures can be a key factor in preventing damage that shortens asset lives.

Exceeding design limits in operation (i.e., using equipment outside of the design envelope) also has a detrimental impact on asset life. For example, electric motors are particularly prone to premature failure if their design limits are exceeded; deterioration of the insulation on windings can occur if the maximum operating temperature is exceeded by even a small margin.

External Factors

M&E assets are subject to external damage (asset strikes, etc.) and damage relating to human error (e.g., cumulative impact of incorrect operation or maintenance procedures).

3.7.2 Factors that Influence the Economic/Service Life of M&E Assets

General Comments

M&E assets can be become uneconomic to run due to the level of reliability. M&E assets can also be replaced because they have become obsolete or because newer equipment is more energy efficient. Changes to process requirements can also require M&E equipment to be replaced, for example, the performance of a pump may be insufficient to meet the demands of an upgraded treatment process.

Issues relating to obsolescence (spares availability, technical support and changing regulatory requirements), as well as the economics of the sector, mean that extending the life of an M&E asset is not always desirable, even when technically feasible. Where an authority owns a number of similar aging assets, spares can be scavenged from other assets to extend life, but this is a self limiting strategy. Building in appropriate spares holdings strategies at the start of the asset's life cycle can thus extend the life of assets.

Service/Performance Measures

Common KPIs used for M&E assets are based on measures of asset reliability. For example, mean time between failures (MTBF) is calculated by dividing the operating time by the number of failures within a specified period. The definition does not distinguish between repairs to different components, or the extent of the repair. As such, MTBF must also be considered in terms of the overall maintenance costs associated with the asset. Other indicators relating to the process or energy efficiency of the asset can also be useful for some asset types; for example, pumps and blowers. In summary, KPIs of use include:

- Number of failures of assets/components, and trends in failures over time
- Measures of asset reliability (MTBF, availability, reliability, etc.)
- Maintenance expenditure (cost of spares, maintenance man hours, call outs cost, etc.)
- Trends in or measures of power/energy consumption
- Measures of process efficiency

Poor performance of an M&E asset or group of assets measured through these KPIs will influence the decision to replace or rehabilitate assets, and thus the issue of remaining asset life.

3.8 ICA Assets

Instrumentation, control and automation (ICA) assets are the hardware and software that allow control and monitoring of processes within pumping stations, treatment works, etc. This type of asset includes online instrumentation (level, temperature, pressure, flow, dissolved oxygen, turbidity and other water quality sensors, etc.), and for the purposes of this research are taken to include telemetry and SCADA assets (consisting of microprocessor; displays and signal input/output facilities, cabling etc.).

Characteristic features for end of asset life for ICA assets are summarized in Table 3-8, and factors that influence asset life are summarized in Table 3-9 and discussed in the sections that follow.

Table 3-8. Characteristic Features of End of Life for ICA Assets.

Asset Category	Characteristic Features of End of Life
ICA	Excessive number/frequency of failures/breakdowns. Excessive maintenance costs being incurred. Electrically unsafe. Obsolete Inoperable

Table 3-9. Factors that Influence the Life of ICA Assets.

Category	Factors
Design and Installation	Manufacturing quality Inappropriate design of components (specification or design details) Asset protection (environmental protection standards with respect to moisture, dust, etc.) Installation method Standard of workmanship
Operating Environment	Exposure to elements Appropriate levels of air conditioning, heating, air filtering, etc. Ambient temperature Humidity Sewage characteristics (including strength, presence of foam, fats, or salts, etc.) Sludge characteristics Water damage Operating temperature Over-current or over-voltage events (i.e., higher current/voltage compared to rating) Power supply (variations, interruptions, etc.) General housekeeping (items free of dust, dirt, etc., as far as is practicable
Operation and Maintenance	Incorrect maintenance procedures Level of preventive maintenance (inspection and diagnosis of faults) Control relays chatter
External Factors	External damage Human error (e.g., cumulative impact of incorrect operation or maintenance procedures) Calibration Storm events (wind damage and lightening)
Economic/Service Factors	Regulatory changes and/or changes to operational requirements (including capacity issues) Technical innovation and obsolescence (e.g., due to spares availability, lack of technical support, etc.) Level of asset redundancy Cost of repair, rehabilitation or other options Stakeholder requirements with respect to monitoring Available budgets Level of service and risk measures

3.8.1 Factors that Influence the Physical Life of ICA Assets

Design and Installation

ICA assets are generally made up of a range of components. Individual components may not be repairable, but overall the units generally are. Sudden early life failures may occur for example where there are "hidden" manufacturing defects or a component has been incorrectly designed or specified. Detail design of the assets at the component level thus impact the expected life of the overall unit.

ICA assets are often configured to provide redundancy and backup so that performance can be maintained if an item of equipment fails. The level of redundancy built in at the design stage can have a significant impact on the level of risk than can be accepted, and thus on the concept of remaining asset life.

Operating Environment

Some ICA components, for example ultrasonic level detectors in pumping station wells, are subject to aggressive operating environments, whereas many ICA assets will be housed in relatively clean areas (for example, in dry wells).

Deterioration processes depend on the type of component; for example:

- Breakdown of insulation can occur due to heat.
- Deposits and corrosion can lead to deterioration of sensing elements.
- Corrosion of electrical contacts and deterioration of interconnecting cabling can occur due to the operating environment (moisture and temperature) and self-generated heat.

Operation and Maintenance

As with M&E assets, correct maintenance and operation of ICA assets can extend life significantly, especially in comparison to the expected life if operation and maintenance are incorrectly carried out. Upgrades to software can also extend the life of ICA assets.

Inappropriate response to minor faults, like ignoring alarms and leaving cooling fans inoperable, can greatly reduce the expected life of an ICA asset. Simple preventive maintenance tasks will extend asset life; for example, checking instruments for damaged bodies and corrosion, checking that electrical conduit connections are complete and intact, and checking wiring terminations for tightness and corrosion.

External Factors

Telemetry connections over telephone lines are subject to damage by extreme weather events, etc. ICA assets are also vulnerable to lightening discharges unless appropriately protected.

3.8.2 Factors that Influence the Economic/Service Life of ICA Assets

General Comments

Factors that influence ICA assets are similar to those for electrical assets, so reliability and functional considerations are important. Additionally, because of the IT component of these assets (software and hardware), ICA systems are especially subject to technical and other obsolescence.

Service/Performance Measures

As with M&E assets, common KPIs used for ICA assets are based on measures of asset reliability and the overall maintenance costs. In summary, KPIs of use include:

- Number of failures of assets/components, and trends in failures over time
- Measures of asset reliability (MTBF, availability, reliability, etc.)
- Maintenance expenditure (cost of spares, maintenance man hours, call outs cost, etc.)

Poor performance of an ICA asset or group of assets measured through these KPIs will influence the decision to replace or rehabilitate assets, and thus the issue of remaining asset life.

3.9 Research Challenges: Asset Life Concepts

From the information presented in this chapter, the challenge associated with assessing remaining life can be summarized thus:

1. Work should be undertaken to develop further the list of factors detailed for key asset types
2. Work should also be undertaken to identify the most significant factors using analysis of data, expert opinion, industry reviews, etc.

CHAPTER 4.0

THE INFLUENCE OF RISK ON REMAINING ASSET LIFE

4.1 Summary

This chapter outlines the importance of risk to asset management and the concept of remaining asset life. Risk is considered to be the product of the likelihood and consequence of failure.

Likelihood of failure can be conceptualized in terms of a simplified load-capacity relationship. Assets fail when the load exceeds the capacity of the asset to resist the imposed load. In this context, 'load' is taken to be any measure of stress imposed on an asset or system, and 'capacity' is the ability to function under the imposed stress. As such, capacity can relate to the strength of the asset or a more complex measure of capacity, such as the capacity of a treatment process. Many assets deteriorate, so the capacity of the asset to resist loads decreases over time. Similarly, asset loading can vary. Final failure is often associated with a particular over-loading event, and failure probability can be considered in terms of the probability of the over-loading event occurring.

Many failure consequences can be reduced to financial costs, though others are more intangible. The determination of whether an asset is at the end of its life is significantly influenced by whether or not the analysis considers only those costs internal to the water authority, or whether a more holistic consideration of cost and consequences is taken.

The potential consequences of asset failure can be used to give a measure of asset importance, which in turn can be used to specify the management strategy adopted for the asset, including the time at which the asset is considered to be at the end of its life. In particular, reactively managed assets will tend to have end of life dictated by failure events, while proactively managed assets will have end of life dictated by other measures of asset deterioration such as condition or performance.

Risk can be considered a cost in the same way as other capital and operational costs, so approaches like Life Cycle Costing or similar can be used to determine if it is more economic to retain an existing asset or renew it in some way. Calculating the net present cost of both retaining and replacing the asset, taking into account the benefits of capital deferral and future failure avoidance, provides a direct means of assessing which option is the most economic. In practice, for many wastewater assets, this analysis often involves determining the optimal time to make an intervention. The optimal time for replacement is when the total expected cost is minimized. In practice, it may be more practical to attempt to maximize the likelihood of successful future performance at reasonable cost, rather than minimize cost explicitly.

4.2 Introduction

Various generic factors were outlined in Chapter 3.0 that influence the time at which an asset is replaced. Many of the factors relate to asset failure, the likelihood of failure, or the cost and other consequences associated with failure. As will be shown in this chapter, the influence of these factors can be considered within a risk-based management framework. Risk is therefore an important consideration in the assessment of remaining asset life, and this chapter addresses the following aspects of this subject:

- The conceptual importance of risk to asset management.
- The likelihood-side of risk, focusing on a conceptual model of load-capacity; load is taken to be any demand/stress imparted on an asset and capacity is the asset's ability to withstand the load and still provide service.
- The consequence-side of risk; including consideration of various cost categories.
- The use of asset importance (assessed in terms of potential failure consequences) to determine the maintenance strategy adopted for the asset, and ultimately the time at which the asset is considered to be at the end of its life.
- The impact of risk on end of asset life decisions.
- A summary of anticipated research challenges arising from the information presented.

This chapter is based on conceptual arguments and descriptions, rather than an explicit consideration of the way in which failure likelihood and consequence are assessed in the context of remaining asset life. A more tangible consideration of tools and approaches used is given in Chapters 5.0 and 6.0.

4.3 An Overview of Risk in Asset Management

Risk is often considered to occur when two or more outcomes are possible and at least one of the possibilities is undesirable. Hence, risk involves both uncertainty (if an outcome is certain or has already occurred there is no risk) and the potential for an adverse consequence (although risk concepts can also be used to consider positive consequences). In asset management, risk is often defined as the product of the likelihood of a failure or other event that will cause loss, and the severity of the consequence (i.e., the magnitude of the loss, Faber and Stewart, 2003; ABS, 2004; IPWEA, 2006). Hence, risk is expressed as:

Risk = Likelihood (either probability or frequency) x Consequence

From this relationship, it can be seen that risk is decreased by either reducing the likelihood of failure and/or limiting the potential consequences of an incident/event. Asset managers thus need to understand both the likelihood-side and consequence-side of risk.

4.3.1 Risk Cost

As shown above, the likelihood-side of risk can be expressed either as the *probability of an event occurring* or the *frequency with which the event occurs* (or is expected to occur). Using frequency as a measure of likelihood leads to a useful conceptualization of risk. If frequency is expressed in units of "per year" and consequence in dollar terms, then the expected cost of risk is given in units of $/yr and can be considered in the same way as any other annualized cost. In particular, the expected cost of risk can be used in cost-benefit analysis of different asset management strategies. The expected cost of risk can be referred to as 'risk cost'. For example, the International Infrastructure Manual (IPWEA, 2006) defines a 'risk cost' as the assessed annual cost to the consequence of a [risk] event; risk cost being equal to the costs relating to the event multiplied by the probability of the event occurring (as per the definition of risk given above).

While there is a natural tendency to give more emphasis to failures that occur more often, or that occur with bigger consequences, it is **risk cost** (i.e., the combination of likelihood and cost consequences) that should be used in asset management decision making. To illustrate why this is the case, consider an asset that causes $100,000 losses per failure compared to an asset that causes $20,000 losses per failure. The asset causing the greater loss per failure would seem to be the most problematic. However, if the first asset fails only twice per year, and the second asset fails 10 times per year, each asset results in $200,000 losses each year. The economic impact associated with operating these two assets is therefore equal.

4.3.2 The Importance of Risk to Asset Management

When making asset-related decisions, a water authority must consider not only the impacts of those decisions on its financial performance, but also on the level of service provided to the community and the environment, and the "value for money" that level of service represents (Buckland, 2000). A fundamental part of achieving "value of money" is the identification and quantification of the various asset-related risks. In fact, it can be argued that asset management is a formalized approach to risk management of physical assets (e.g., Burn et al., 2007).

Essential components of a successful asset management methodology are the ability to predict the time-dependent deterioration and failure of an asset, and techniques to quantify the consequences that are incurred upon failure; in other words, to calculate time-dependent risk (Davis et al., 2008). The level of risk associated with a particular asset can then be constrained and used to schedule future interventions. In this respect, capital renewal works can be considered as replacement investment undertaken to reduce operations costs, maintenance costs and/or enhance performance (Davies et al., 2001a). From this perspective, the aim of a successful asset management methodology is to achieve an optimal (or at least acceptable) balance between intervention costs (including repair and replacement) and cost of asset maintenance and operation, through the consideration of risk and other costs involved.

The need to balance risk-costs against other costs can be illustrated by considering the theoretical balance between total intervention costs (inspection, maintenance, and renewal) and risk-costs associated with asset failure. For example, in the literature relating to the maintenance of manufacturing and process equipment, plots such as that shown in Figure 4-1 are often used to demonstrate that there is a trade-off between the levels of maintenance, the corresponding risk-costs associated with equipment failures, and total cost (e.g., Woodhouse, 1999). Similar plots have been produced that show the trade-off between risk costs and the cost of pipe-asset renewals (e.g., Shamir and Howard, 1979), and the cost trade-offs involved in sewer operation, inspection and rehabilitation (e.g., Madryas and Przbyla, 1998).

As shown in Figure 4-1, it is theoretically possible to balance maintenance expenditure against risk-cost such that the overall cost of asset ownership is minimized. Importantly, increased reliability (reduced risk-costs) can be achieved through higher maintenance expenditure, but only by accepting an increase in total costs. In purely financial terms, the increase in expenditure to achieve this level of asset performance would not be justified. While in practice it may be difficult to undertake analysis to develop the type of curve shown in Figure 4-1, reducing risk to an economic/acceptable level through a judicious combination of maintenance and renewal activity is still a fundamental aspect of asset management (Burn et al., 2007). The interpretation of these concepts in terms of asset replacement is discussed further in Chapter 5.0.

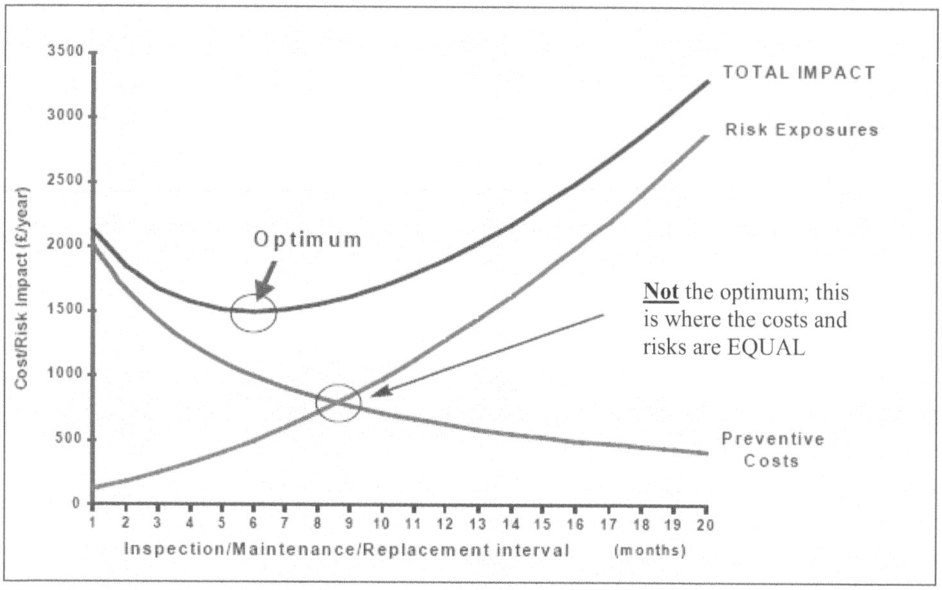

Figure 4-1. Optimal Timing of Intervention.
(Reprinted with permission from Woodhouse, 1999)

4.3.3 A Life Cycle Focus

As well as balancing risk and other costs, another broad objective of asset management is to provide asset capacity that allows a water authority to deliver sustainable levels of service while meeting financial, health, safety, environmental, and other regulatory and non-regulatory obligations (e.g., WERF, 2007). A key part of this challenge is the optimal utilization of the remaining life of existing assets (Schneider et al., 2006). At some point, however, it will be more economical to replace an existing asset. A key requirement of asset management is thus to achieve an appropriate balance between utilizing existing assets and investing in renewals. In turn, achieving this balance requires that asset management decisions be considered from an asset life cycle perspective.

Schneider et al. (2006) indicate that a comprehensive approach for asset management focuses on the life-cycle costs of the individual assets that make up the system. However, in practice a life cycle approach to asset management must also recognize and account for the many inter-dependencies between assets, their capability, the various stages of asset life cycle and the associated life cycle costs, as illustrated schematically in Figure 4-2 (after Buckland, 2000). As suggested in Figure 4-2, a life cycle focus requires various cost elements to be taken into consideration; for example:

- Capital cost of the original asset ($)
- Maintenance cost ($/yr)
- Failure and repair costs (i.e., the cost of risk in $/yr, which can include consideration of external costs and externalities)
- Replacement cost ($)
- Operations cost ($)

- Disposal costs and any resale value ($)

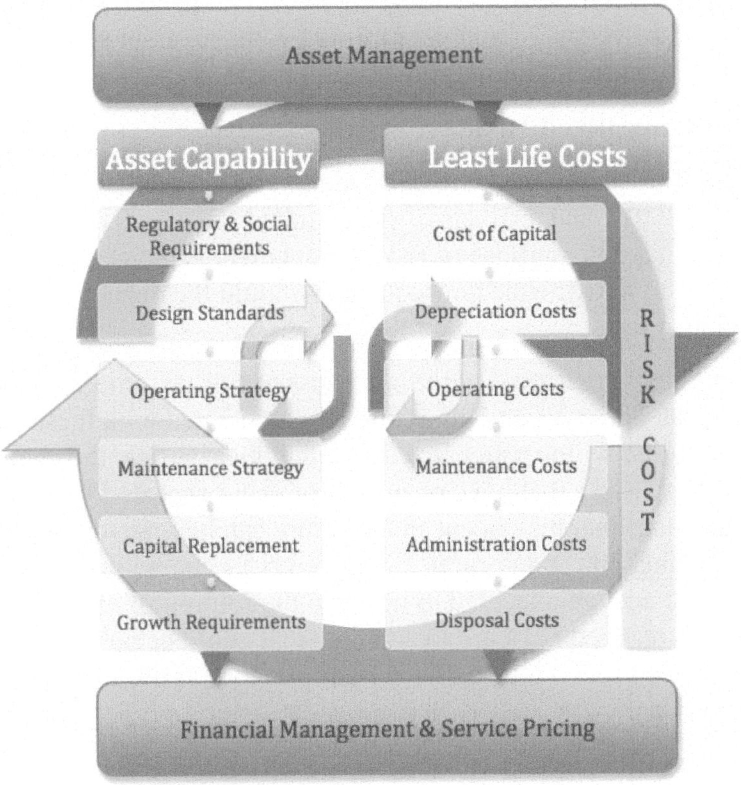

Figure 4-2. A Life Cycle View of Asset Capability.
(Reprinted with permission from Buckland, 2000)

Various approaches have been developed for the analysis of asset life cycle costs (e.g., Boussabaine and Kirkham, 2004), each of which differ in the time scales considered and scope of the costs/impacts analyzed. For example, in life cycle costing (LCC) the total costs for an asset borne by the water authority are estimated over some defined period. Whole life costing (WLC), on the other hand, considers the whole asset life, and includes cost elements for which the water authority is directly responsible and external costs borne by others. In both cases, since costs are incurred at different times, it is necessary to take into account the time value of money through a depreciation term. In practice, assessment of life cycle costs is associated with significant levels of uncertainty, so any measure of life cycle cost is a random variable and its expected value serves as a criterion for decision making (Val and Stewart, 2005).

A key component of any LCC or WLC is the cost of risk. As indicated in Figure 4-2, risk costs can arise from each stage of the asset life cycle, and asset solutions should be selected in light of these costs. At the design stage, life cycle costing and/or risk analysis techniques can be used to select an appropriate solution, taking into account the expected failures over the asset's life (Faber and Stewart, 2003; Val and Stewart, 2005). Once constructed, the management of failure risk requires Authorities to make ongoing assessments of the level of risk, and to determine an appropriate risk management strategy that represents the balance between the assessed level of risk and other life cycle costs, including the cost of asset rehabilitation or replacement.

In summary, risk has a significant influence on decision making throughout the life cycle of an asset and system of assets, including determining when an asset should be replaced. Risk is therefore a key consideration in defining remaining asset life.

4.3.4 Realized and Unrealized Risk

Risk can be considered as a measure of the degree of exposure to probabilistic loss events (Buckland and Hasting, 2001), and analysis of risk is inherently concerned with understanding future (probabilistic) events. The purpose of risk management is then either to prevent these loss events from occurring or to mitigate the consequences should the event occur (e.g., IPWEA, 2006; Moglia et al., 2006).

While risk analysis is concerned with future loss events, as time goes by, some of these events will actually occur. The costs and consequences associated with these failures (or other loss events) are then incurred by the water authority and other parties; the event is no longer a risk, since it has happened. Risks that have occurred (or occur repeatedly) can be referred to as ***realized risks***, whereas those that may occur in the future can be referred to as ***unrealized risks*** (e.g., Buckland and Hasting, 2001). In practice, realized risks can be future risks of a kind that have happened in the past and are expected to occur in the future, for example routine failure of a small diameter sewer. Unrealized risks are loss events that could occur, irrespective of whether they have ever actually occurred in the past. For example, loss of life associated with a sewer collapse under a busy road.

As shown in Figure 4-2, risk costs form part of the asset life cycle cost structure. However, since risks can be realized or unrealized, historical cost may or may not reflect the level of risk exposure associated with an asset. In particular, historical cost records will reflect the cost of routine failures at least to some degree. In contrast, significant unrealized risk may not be reflected in historical records at all. It is important to consider all risk costs, including those that are unrealized, when making asset management decisions (e.g., Buckland and Hasting, 2001).

4.4 Likelihood of Failure

In general, it is possible to categorize factors that influence likelihood of failure into those that determine or influence the ***capacity*** of an asset to resist loads and those that determine or influence the ***load*** imposed on the asset. For example, De Silva et al. (2002) note that ferrous pipes (such as cast iron) can fail in service due to pitting corrosion (which reduces structural capacity) and excessive in-service loading. For such assets, in-service loads can be split into contributions from internal pressure (if any) and external soil loading attributed to seasonal ground movement and traffic loads. It follows that failure will occur when these loads combine to exceed a critical property (i.e., capacity) of the corroded pipeline, such as its yield strength or critical strain energy.

In this example, the capacity of the asset is characterized by a measure of the pipe's strength and the load is a composite structural loading. The details of the load and capacity for other asset types, and the factors that influence them, will vary, as outlined in Chapter 3.0. These details are an important consideration in determining remaining life of a given asset. For the purposes of the current discussion, however, it is sufficient to recognize that the probability of failure can be idealized in terms of a simplified relationship between 'load' and 'capacity'.

Again, for the purposes of this discussion, 'load' can be taken to be any measure of stress imposed on an asset or system, and 'capacity' the ability of the asset to function under the imposed stress. An obvious application is in the case of structural analysis, where 'load' is the stress imparted on a structural element and 'capacity' is the ability of the element to resist that load (as noted above, expressed in terms of some critical measure of the asset's strength). Load and capacity concepts can also be used for less obvious applications; for example, the quality of inflow to a wastewater treatment works can be considered a load, with capacity being a measure of the work's ability to treat the wastewater adequately. Similarly, load and capacity relationships can be applied to 'natural assets' such as receiving waters; for example, where load is a measure of the effluent discharge and capacity is a measure of the assimilative capacity of receiving water (e.g., Ganoulis, 1994).

With these definitions in mind, it can be asserted that for successful operation or provision of service, the load (L) imposed on an asset (or system) must be within its capacity (C). As discussed in Elishakoff (2005), a common engineering approach for expressing the capacity of an asset to resist imposed loads is in terms of safety factors (SF). In simple terms, a safety factor can be defined as the ratio between some measure of load and an equivalent measure of capacity (e.g., the ratio between asset strength and an equivalent measure of imposed load or stress). At any time in its life cycle, an asset's safety factor can be expressed in terms of the ratio of capacity (C) to load (L); that is:

$$SF = C/L$$

The likelihood of failure can then be written as:

$$\text{Likelihood of failure} = \text{Probability}\ (SF<1) = \text{Probability}\ (C-L< 0)$$

In other words, the asset fails when the load exceeds the capacity, which implies that the factor of safety is less than unity (i.e. SF<1).

The expression C-L<0 is a form of function referred to as a 'limit state function' (e.g., Cheung and Kyle, 1996; De Silva et al., 2002; ABS, 2003). In terms of remaining asset life, as a limit state function approaches zero (which is equivalent to the safety factor approaching 1), an asset can be considered to be approaching failure, and this in turn could be taken as an indication that the asset is approaching the end of its life. A safety factor less than 1 implies the asset is in an unacceptable state. By implication, and depending on the measure of load used (e.g., peak or average load), the asset has either failed or the probability of failure is unacceptably high.

In reality, load-capacity relationships are generally complicated, so characterizing the safety factor and probability of failure is a complex and involved task (e.g., see Davis et al., 2007; De Silva et al., 2006; Faber and Stewart, 2003). In particular, the capacity (C) and load (L) are specific to a given failure mode (e.g., Rausand and Øien, 1996), and are comprised of base variables that describe the physical failure mechanism that occur under operational conditions (Davis and Marlow, 2008).

4.4.1 Changes in Asset Load and Capacity

An additional complexity in determining the probability of failure (or factor of safety) is that both load and capacity change over time. Loads can vary due to changes in operating environment and/or operational demands. For example, sewerage assets can be subject to increasing dynamic (traffic) and static (building) loads over the life of the asset, which can eventually contribute to failure. Similarly, electrical assets can be placed under greater loads due to increasing operational demands.

Processes that reduce the capacity of an asset as it ages, such as material degradation, wear and fatigue, are referred to generically as 'deterioration'. The process of deterioration varies according to asset type and materials of construction (e.g., Esselman et al., 1998; Davies et al., 2001a). Deterioration can progress without an asset failing for much of its life, but eventually a critical point will be reached where the load imparted on the asset exceeds that capacity of the asset to resist the load and functional failure then occurs.

Figure 4-3 illustrates schematically how capacity and load change over time. In the Figure, capacity is assumed to decrease gradually over time due to deterioration. Three loading scenarios are also represented:

- The load remains constant over time;
- The load falls over time; and
- The load increases over time.

Given the same rate of deterioration (loss of capacity), each of these loading scenarios has an impact on the time to asset failure (defined as the time at which load exceeds capacity), as illustrated in Figure 4-3. In particular, if the loads increase over time as the capacity falls, failure occurs earlier than compared to the situation where loads remain constant or falls over time (leading to the different failure points f_1, f_2, f_3). The way in which loads change over time can therefore be a significant factor in determining the remaining life of an asset.

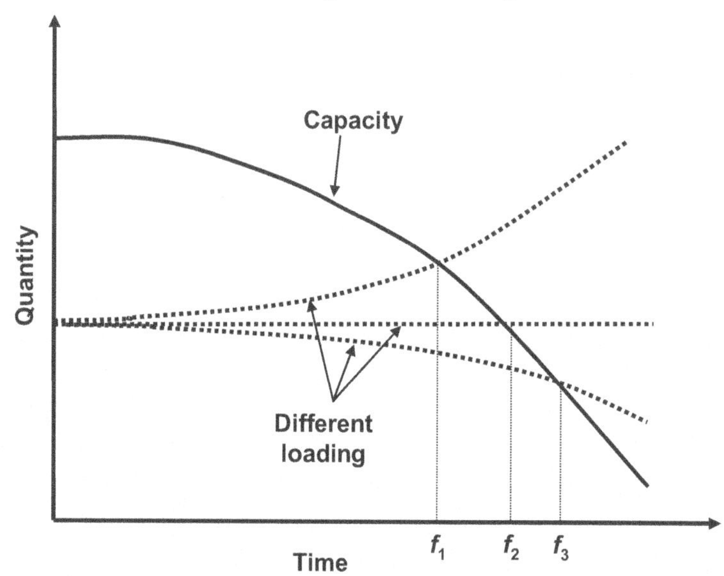

Figure 4-3. Simplified Load-Capacity Relationships.

4.4.2 Event Driven Failures

Figure 4-3 represents a deterministic view of asset failure, since there is an implicit assumption that load and capacity can be meaningfully represented as point values. The time to failure is then defined as well defined points f_1, f_2, f_3. In practice, however, the capacity of similar assets will vary and the capacity of any given asset will be generally uncertain. In addition, the load imparted on the asset varies in both time and space. As such, load-capacity relationships are much more complicated than is suggested by Figure 4-3, and both load and capacity must be considered stochastic variables (e.g., Cheung and Kyle, 1996; Sadiq et al., 2004; Melchers, 2005; Davis and Marlow, 2008).

With these issues in mind, Figure 4-4 shows a more realistic (if still simplified) representation of load-capacity relationships that reflects the stochastic nature of the variables. As shown in Figure 4-4, the values of load and capacity are now represented by a distribution about a mean. One implication of the stochastic nature of load-capacity relationships is that failure of an asset does not occur simply as the result of a steady and progressive deterioration to a point where the remaining capacity is no longer sufficient to resist in-service loads (as suggested by Figure 4-3). Instead, failure is the result of a combination of deterioration in condition and the influence of operational or environmental factors that cause peak loads (e.g., Davies et al., 2001; Sadiq et al., 2003; Melchers, 2005).

From this perspective, the likelihood of failure can be considered in terms of the likelihood of observing an overload event (i.e., when load exceeds capacity), which again will vary over time. For example, consider the situation illustrated schematically in Figure 4-4. The area of operation over which an overload event can occur is represented by the shaded area. Hence, at time t_1 there is only a small overlap between load and capacity. In comparison, at time t_2, the average load has increased and the average capacity has decreased, though the distribution of load and capacity around the mean are unchanged in this case. From visual examination, it can be seen that, in relative terms, the likelihood of failure at time t_2 is greater than at time t_1. The absolute likelihood of failure will depend upon the probability of an overloading event occurring.

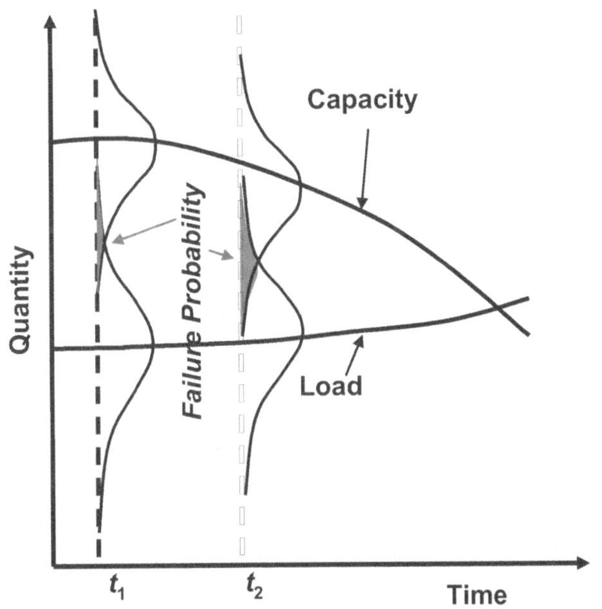

Figure 4-4. A More Realistic Load-Capacity Relationship.

As suggested in Figure 4-4, the probability of failure tends to increase with time as the asset deteriorates because of the change in load-capacity relationships. However, while it is useful to illustrate the concept of failure probability in terms of load, capacity and overload events, it is important to note that the uncertainties associated with the asset's properties, loads, load events, and environmental conditions mean that failures occur as apparently random events with time-dependent probabilities of occurrence (Val and Stewart, 2005).

4.4.3 The Influence of Repair on Failure Probability

For non-repairable assets, failure marks the end of asset life, as discussed in Chapter 2.0. In the case of repairable assets, however, capacity and function can be reinstated through some kind of repair, including the replacement of components (e.g., Cheung and Kyle, 1996; Vanier, 2001). While a successful repair will return an asset to a functioning state, the impact on the subsequent failure probability will depend on the type and quality of the repair undertaken. Most repairs can be expected to reduce the failure probability (in comparison to the probability immediately before failure). However, if a short-term or poor quality repair is undertaken, the failure probability could increase. The asset will still function following the repair, but there is a relatively high likelihood of further failure.

As indicated previously, the future probability of failure is an important consideration in determining if an asset is considered at the end of its life. In particular, if the repaired asset is likely to have poor reliability into the future, an asset could be considered for replacement. Reliability is defined as the probability of a product performing its intended function under stated conditions without failure for a given period of time; reliability is thus the probability of non-failure over time (e.g., Ebeling, 1997). Reliability can also be considered to be the probability of survival of a structural element or component (Cheung and Kyle, 1996). The relationship between reliability and probability of failure can thus be conceptualized as follows:

Reliability = Probability of survival = (1 – Probability of failure)

Schneider et al. (2006), present a conceptual model of the aging cycle for an asset based on reliability considerations, as shown in Figure 4-5.

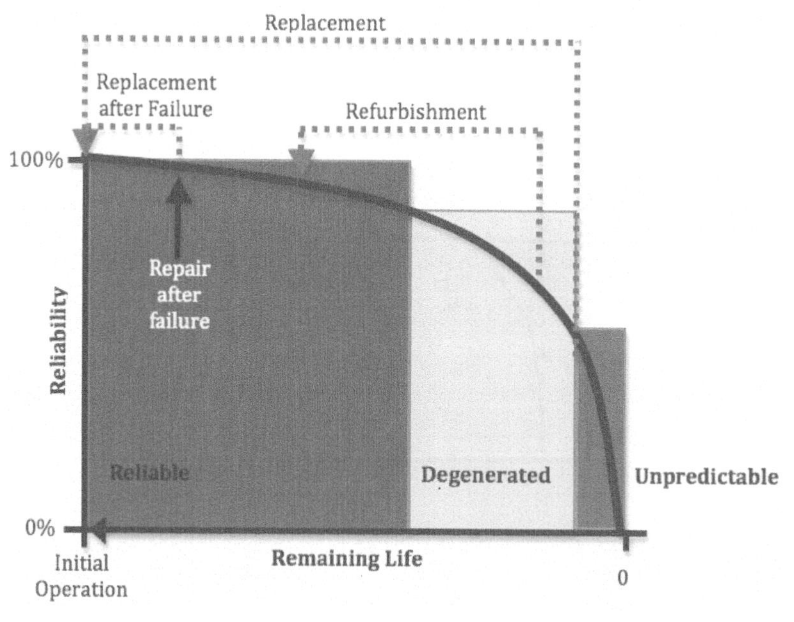

Figure 4-5. The Impact of Repairs on Asset Reliability.
(Reprinted with permission from Schneider et al., 2006)

With reference to Figure 4-5, the conceptual model describes an asset with three different states: reliable, degenerated and unpredictable. During its lifetime, an asset will pass through each state and spend a certain amount of time in each. Maintenance and refurbishment will slow down the rate at which an asset moves from one state to the next. In each state, an asset will have a different level of reliability. Repair and refurbishment have the effect of improving the reliability of the asset, though an 'as new' state may not be achieved. Replacement of the asset is eventually undertaken when the existing asset is considered to be unreliable and economic refurbishment is no longer an option.

While reliability is a concept that is familiar to asset managers, analysis of risk involves explicit consideration of failure probability. It is thus worth illustrating how the probability of failure for an asset could vary over time as it is repaired and refurbished. To this end, Figure 4-6 mirrors the scenario illustrated in Figure 4-5, but shows how probability of failure would change rather than reliability. With reference to Figure 4-6, it can be seen that the probability of failure of the asset initially increases as it deteriorates. At time t_1, a failure occurs and the asset is repaired. The repair and maintenance work carried out has the effect of resetting the probability of failure to the 'as new' condition (i.e., the probability of failure of the repaired asset is approximately that of the new asset).

Following the initial repair, the failure probability increases over time until failure again occurs at time t_2. At this point, the asset is refurbished. This has the effect of reducing the probability of failure (compared to that just before failure occurred), but in this case the probability of failure exceeds that of the 'as new' asset. In such cases, there would be a benefit of replacing the asset with a new one, because the expected future cost of failures would then be reduced (see Section 3.6 for more discussion on this issue).

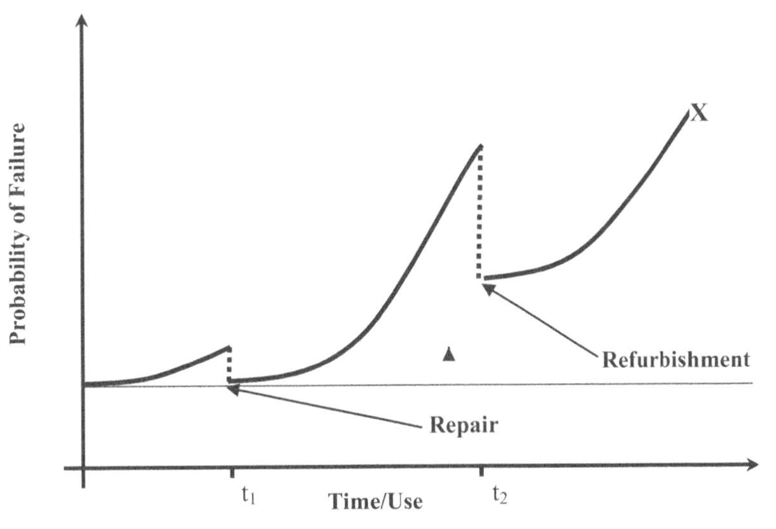

Note: The figure is drawn to highlight the fact that a failure can occur at different failure probabilities and the characteristic increase of failure probability after a repair or refurbishment, is not necessarily the same as that for an 'as new' asset.

Figure 4-6. The Impact of Maintenance on Asset Failure Probability.

As noted in Chapter 2.0, in practice it may be undesirable for an asset to fail. In such cases, the interventions (repair and overhaul) at time t_1 and t_2 could be programmed proactively through condition and performance monitoring, or some other approach like use-based maintenance scheduling (see WERF, 2007). Whether or not an asset is managed using a proactive or reactive maintenance approach depends upon the consequences of asset failure, which are considered in more detail in Section 4.5.

4.4.4 Influence of Failure Modes

Assets can generally fail in a number of different ways, and the way in which an asset fails to fulfill its function can be referred to as its ***failure modes*** (e.g., Rausand and Øien, 1996). When considering the likelihood-side of failure it is necessary to analyze failure modes separately because each failure mode will have a different limit state and could have different variables.

For many civil assets, public safety is a paramount concern, so analysis of failure likelihood focuses on determining if there will be a catastrophic failure with safety implications. However, as noted by Ellingwood (2007), public safety is a necessary but not sufficient performance requirement. The ability to provide service is a more generally applicable requirement for infrastructure assets, so it is important to minimize the likelihood of business interruption, environmental impact or economic loss. Hence, as well as considering health and safety requirements, the analysis of likelihood should also consider all failure modes that impact service provision.

4.4.5 Acceptance Criteria for Failure Likelihood

Whether or not an asset operates without problems or is subject to failures is obviously of concern and relevance to an asset manager, and the number of failures that occur can thus be used in decision making; for example, a small diameter sewer may be replaced if it becomes blocked 3 times in 5 years (see Section 5.5 on Sydney Water). This concept can be extended to the consideration of probabilistic failures (i.e., unrealized risk, see Section 4.2.4). For example, Ellingwood (2007) refers to the use of acceptable failure frequency in the nuclear industry and gives the example that the total core damage frequency must remain below 10^{-4}/yr. This failure frequency represents a return period of 1 in 10,000 years, so self evidently the acceptability or otherwise of a system to which this criterion is applied has to be judged in probabilistic terms.

While the level of risk is very different in the wastewater industry, acceptance criteria for a given asset type could still be specified in terms of failure likelihood or reliability. However, specifying acceptance criteria in this way does not provide a consistent 'currency' from which to judge the business case for intervening in one asset, when compared to another. As noted in previous sections, in asset management decision-making the fundamental consideration is not acceptable reliability or likelihood of failure, but acceptable risk. This requires the consequence side of risk to be considered either explicitly (e.g., in the calculation of risk-cost) or implicitly (e.g., specifying acceptable likelihoods for a given consequence), as described in the next section.

4.5 Consequence of Failure

By definition, the likelihood of failure dictates how probable it is that an asset will fail, or how often the asset fails. The other aspect of risk is the analysis of failure consequence. From the perspective of service provision and risk-costs, important aspects of this analysis are the need to understand:

- Tangible and intangible consequences of asset and service failure;
- The extent of impacts (for example, number of customers effected);

- The severity (whether service loss is total or partial);
- The duration of service disruption (how long customers feel the impact); and
- Generic factors considered in the analysis of the consequence-side of risk.

Table 4-1 categorizes important aspects of failure consequences in more detail.

Table 4-1. Consequence Categories.

Category	Description
Service failures	Operational impacts such as service interruptions, quality or quantity variations, poor customer service, etc.
Customer impacts	Impacts of interruption and customer dissatisfaction,
Direct costs	Including damage to property and infrastructure, operational impacts, increased maintenance costs (associated with repair, excavation, material and equipment, traffic management, and other costs), business interruption, fines, and compensation for flooding (internal and external), etc.
Environmental impacts	Unacceptable environmental impacts considering natural and constructed environments
Externalities	Traffic disruptions, third party discomfort, aesthetic impacts, health impacts
Health and Safety impacts	Possible death or injury to water authority staff and members of the public
Public relations impacts	Public relation issues arising from asset failures

4.5.1 Cost Consequences

Many of the consequences given in Table 4-1 are, or can be considered to be, a financial cost. Some of these costs are ***tangible*** in that they have a market value and can be easily expressed in financial terms, while some are ***intangible***. Intangible costs do not have a market value and are associated with less tangible aspects of failure such as image loss, social impacts and environmental damage (Buckland and Hastings, 2001). While intangible costs do not have an explicit market value, it can still be possible to attribute a monetary value to them (e.g., Marlow and Burn, 2007).

Some of the costs are ***direct costs*** borne by a water authority, such as the tangible costs shown in Table 4-2. Other costs are ***indirect costs*** in that they are not borne directly by the water authority, but are instead borne by customers or communities; for example, the cost of commercial disruption (at least when the cost is not passed on to the water authority in some way).

Table 4-2. Direct (water authority-borne) Cost Categories.

Category	Description
Costs associated with asset failure	Costs of responding to the failure, including any direct financial penalties Costs of restarting, testing and monitoring failed equipment Costs associated with health and safety incidents Disposal of any non-compliant waste
Costs associated with a loss of service	Any compensation to customers (domestic and commercial) Cost of operational responses (e.g., clean-up costs, additional costs associated with effected customers, etc.) Regulatory costs

Other costs are considered to be pure externalities. In economics, the term 'externality' is used to refer to any impact caused by human activity whose costs (or benefits) are not factored into the decisions of the economic agents involved in the transactions (e.g., van Bueren and

Hatton MacDonald, 2004). Put more simply, an externality is a cost borne or benefit received by a third party not involved directly in a commercial transaction (Speers et al., 2002). For example, road disruption caused by a collapsed sewer asset would be considered an externality.

At present, consequence calculations made for asset management purposes do not often include a methodology for factoring in costs external to the water authority (indirect/intangible costs) or externalities (e.g., environmental damage and social costs). However, if external risk-costs and externalities (tangible or intangible) are not given due consideration, the analysis will underestimate the economic value of risk management strategies, including the benefits of replacing or rehabilitating an asset. This in turn can lead to an asset being retained past its true economic life. In contrast, inclusion of external costs and externalities in the analysis of end of asset life reflects the true cost of service provision to the community, and assets are replaced when the overall cost of service provision is minimized (or at least reduced) from the community's perspective (Buckland, 2000; Burn et al., 2007).

Note: the inclusion of external costs and externalities in the analysis of risk will effectively shorten the economic life of an asset by placing a greater value on failure avoidance. Rigorous approaches for the inclusion of these costs are therefore required.

4.5.2 Probability-Side of Consequence

When an asset fails, different failure consequences arise depending upon the specific context of the failure (e.g., Val and Stewart, 2005). Hence, the failure of the same asset under different operating demands and conditions can produce vastly different failure consequences (for example, the failure of a car tire at low speeds often will produce less impact than failure at high speeds). Thus, consequences of failure vary depending on the specific characteristics of the failure event, and this variation can be considered in terms of probabilities.

Ideally, the probability-side of consequence would be considered in risk analysis for asset management purposes, but often this is not practicable given the data and resources available. Instead, assessments are generally made in terms of some assessment of maximum consequence of failure that can be reasonably expected to occur. For example, the fact that an asset could cause loss of life does not mean that a life will be lost when an asset fails, but the management strategy adopted would be selected with this potential consequence in mind. This aspect of asset management is considered in more detail in the next Section.

4.5.3 Consequence as a Measure of Asset Importance

In general, unless there is a major change in the use, service duty or service parameters for a given asset, the potential consequences of failure are likely to remain fixed for its service life (e.g., ABS, 2003). This allows the consequence of failure to be used as a constant measure of asset importance and to specify the management strategy for an asset. For example, the higher the potential consequences of asset failure are, the more important the asset is and the more effort can be justified to prevent failure from occurring.

While various terms can be used to categorize assets in terms of asset importance (e.g., 'critical asset' and 'asset criticality'), this research categorizes assets solely from the perspective of the strategy adopted to manage the asset. Assets can be managed using a 'reactive' strategy, a 'proactive' strategy, or a 'mixed' strategy depending on the relative balance between potential consequences and the cost of risk mitigation. The concepts underlying this categorization are illustrated in Figure 4-7 (e.g., Marlow, 2007; Burn et al., 2007).

As discussed by Buckland (2000) and shown in Figure 4-7, an asset with low consequence of failure is generally managed reactively; i.e., such assets are left to operate until failures start to occur. This approach can be taken because the expected cost of failure does not justify the cost of preventing that failure (though this does not imply that cheap inspection techniques such as operator inspection should not be used where possible; see WERF, 2007). Once an asset has failed (or starts to fail repeatedly), a decision is made to repair and maintain or replace the asset. This decision can be expanded to consider the replacement of similar assets. The decision to retain or replace the asset(s) would include consideration of budget constraints, the economics of continuing to operate the existing asset, the levels of customer service needed, and operational strategies that can be economically implemented to reduce the impact of retaining a failing asset (Burn et al., 2007).

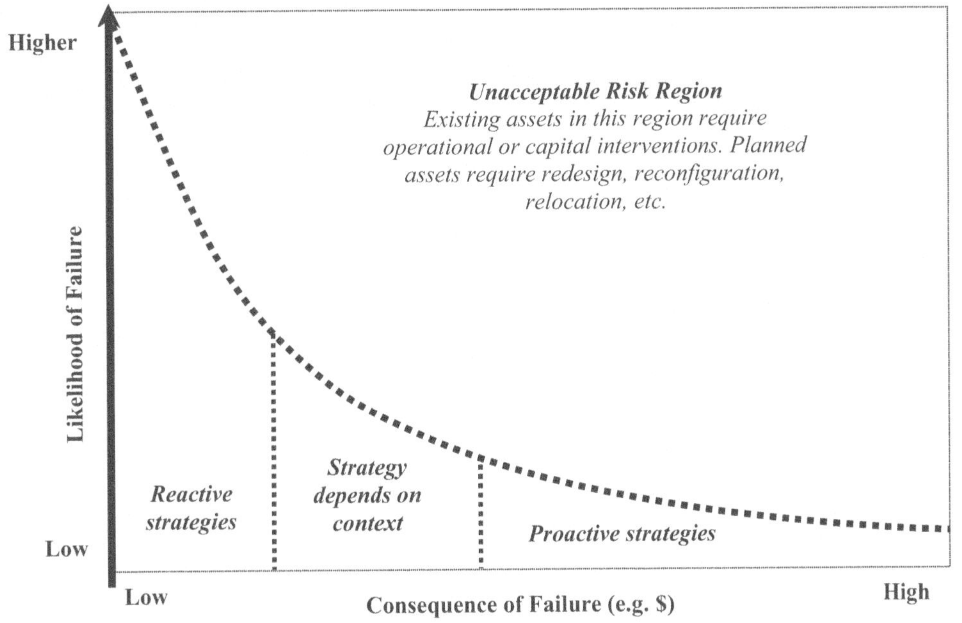

Figure 4-7. Risk Based Management Strategies.
(Reprinted with permission from Marlow, 2007)

As shown in Figure 4-7, proactive strategies generally are applied to assets when the consequences associated with failure are large, and there is thus the potential for Authorities, municipalities and other segments of society to incur high costs (tangible and/or intangible) if failure occurs. For such assets, the economics of preventing failure are advantageous (again see WERF, 2007 for further discussion).

The criteria used to assess end of life are different for reactively and proactively managed assets. In particular, a reactive management strategy is characterized by the use of failure events or failure history in decision-making; individual assets may be run to failure or replaced because of the failure history of similar assets. In contrast, for proactively managed assets, end of life criteria are judged in terms of level of deterioration and predictions of failure, rather than through asset failures *per se*. As such, data on asset condition and performance must be collected and analyzed to determine risk of failure (e.g., WERF, 2007), though early replacement can also be used if the onset of failure cannot be detected economically.

4.5.4 Level of Redundancy

For assets where the main consequence of failure is related to a loss of asset function, one means of mitigating risk is to build redundancy into the system so that when an asset fails (or is taken out of service), function can be maintained by one or more similar assets (e.g., Ebeling, 1997). For example, if a pump operating in isolation were to fail, the pumping function would be lost. If however there was redundancy built into the design (i.e., another pump was operating in parallel, as part of a duty-standby configuration), the pumping function could be maintained should one of the pumps fail. Provided that the standby pump was well maintained and that the pumps were switched at regular intervals, the standby configuration would significantly reduce the risk of service failure.

In practice, redundancy improves the reliability of service provision and reduces the consequence-side of risk associated with asset failure, which translates into lower levels of risk. Building in redundancy, however, comes at additional cost, and this option should thus be considered against other risk mitigation options.

Where redundancy is considered to be a necessary and cost-effective form of risk management, different asset management strategies could be subsequently adopted in light of the reduced asset importance. Whether or not this will have an impact on end of life considerations will depend on the context within which the assets are operating. Nevertheless, a less conservative (less risk-averse) approach to assessment of remaining asset life could be justified.

4.5.5 The Impact of Failure Modes

As noted previously, assets can generally fail in a number of different ways, and the way in which an asset fails to fulfill its function can be referred to as its ***failure mode(s)*** (e.g., Rausand and Øien, 1996). The failure mode of an asset can determine the level of consequence that is generated, and can influence the management strategy adopted and the definition of remaining life. For example, more conservative assessments of remaining life will need to be made for assets that demonstrate catastrophic failure modes (e.g., rupture), in comparison to assets with non-catastrophic failure modes (e.g., pin holing).

4.6 The Impact of Risk on Asset Replacement Decisions

Risk is an important consideration in determining the need for maintenance and, ultimately, is an important consideration in determining when an asset is at the end of its life.

For proactively managed assets, asset management decisions should be aimed at maximizing the likelihood of successful future performance at reasonable, if not minimum, cost (Ellingwood, 2007). For reactively managed assets, in contrast, deciding whether to replace an asset often reduces to determining the trade-off between the cost of retaining the asset, which will include risk costs associated with failure, and the cost of replacement. Nevertheless, whether a reactively or proactively managed asset is considered, the decisions concerning the retention or replacement of an asset involve implicit or explicit consideration of both the costs and the benefits associated with each available option.

Since for many infrastructure assets the service provided must be maintained into the future, the benefits associated with service provision (including the revenue generated) are often approximately the same for the old and new asset. However, replacing an asset can accrue additional benefits associated with avoidance of future failures. This concept is illustrated in Figure 4-8. As shown, replacing an asset means that future failures will be avoided, and these avoided failure costs and service issues can be considered benefits of replacement (see Davis and Marlow, 2008 for more details). It should also be noted that, because of the time value of money, deferring capital investment into the future is financially beneficial.

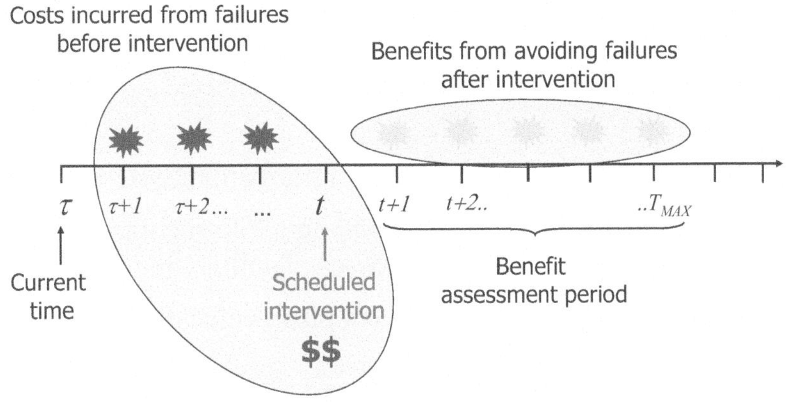

Figure 4-8. Illustration of Intervention Costs and Benefits.
(Reprinted with permission from Davis & Marlow, 2008)

Calculating the net present cost of both retaining and replacing the asset, taking into account the benefits of capital deferral and future failure avoidance, provides a direct means of assessing which option is the most economic (again, see the example given in Davis and Marlow, 2008). In practice, and as suggested in Figure 4-8, for many wastewater assets this analysis often reduces to determining the optimal time for an intervention, since the asset will need to be replaced at some time (e.g., Shamir and Howard, 1979; AwwaRF, 2005b; Davis and Marlow, 2008).

4.7 A Case Study of Risk in Practice: The City of Henderson

This chapter has presented key concepts as they relate to the role of risk in asset management. Some of the concepts require a significant amount of data and analysis to apply fully. Such analysis can be undertaken for specific assets where there is a need to understand risk and remaining life accurately. In practice, however, water authorities also need pragmatic approaches to understand risk across their asset portfolio. Assets associated with high levels of risk can then be analyzed in more detail and/or risk mitigation strategies selected.

The type of risk analysis undertaken at the portfolio level involves the same risk concepts as described above, but these are (by necessity) applied in a more pragmatic manner using available data and expert opinion.

For completeness, a case study has been included herein to show how one U.S. water authority undertook a high level analysis of asset-related risk.

4.7.1 The Department of Utility Services of the City of Henderson

The Department of Utility Services (DUS) of the City of Henderson, Nevada, provides asset management stewardship for an array of assets including 750,000 square feet of maintained facilities, almost 2,000 miles of utility lines and two major water and wastewater treatment facilities valued in excess of $1.5 billion. The city has identified several drivers for improving asset management including 10% annual growth for the last decade, current expansion of city facilities and aging infrastructure, higher community expectations and the need to "do more with less." To address these drivers, the city is undertaking significant initiatives over the next five years including:

1. Development and implementation of an extensive Asset Management program including improved workforce management planning and the ability to electronically dispatch/receive work under information to/from field staff, and the integration of existing asset management applications into the selected asset management software package;
2. Continued enhancements to the Customer Information System;
3. Expansion of the capabilities of the Geographic Information System (GIS) to strengthen the city's ability to manage assets, facilities and land-related information;
4. Preparation of a strategic asset investment plan, which outlines the strategy for managing their physical infrastructure.

A schematic of the city's asset management program is shown in Figure 4-9. Selected benefits of the program include reliable infrastructure due to effective management of life cycle costs, increased efficiency of core services by implementing data management and information technology applications, and investment planning that meet the city's business goals.

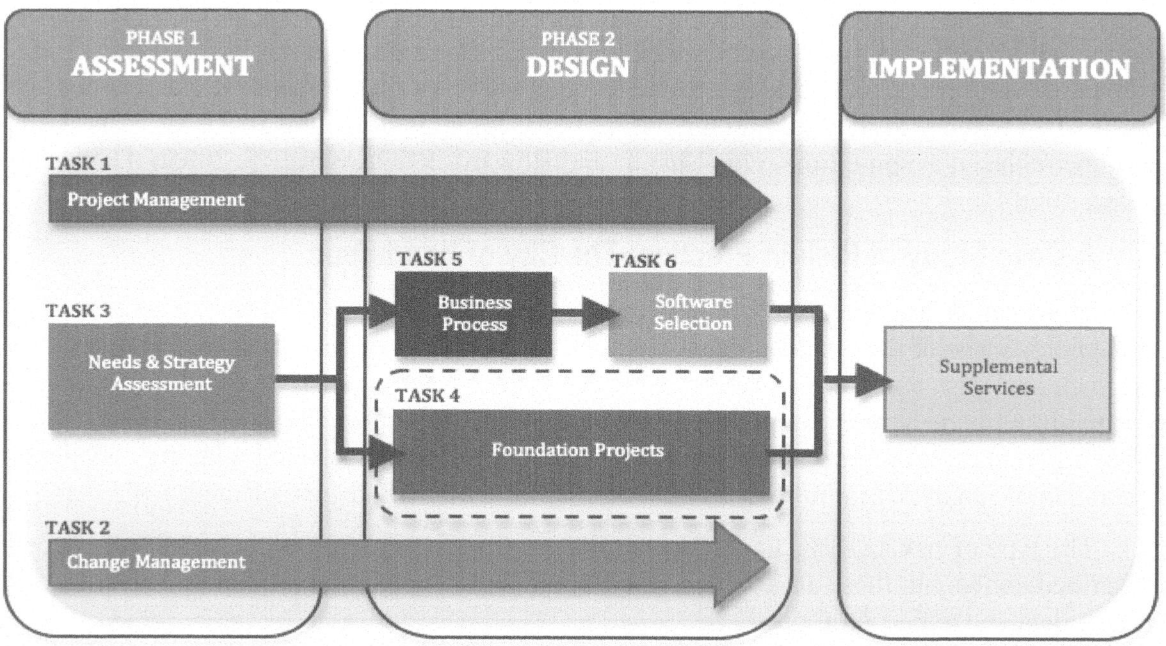

Figure 4-9. City of Henderson Asset Management Program.

4.7.2 Assessment of Portfolio Risk

In order to accelerate the implementation of the asset management program, a high-level risk assessment of the utility's asset portfolio was completed. This project was identified during the assessment phase of the program (Phase 1; Figure 4-9) and considered to be a critical foundation project (i.e., part of Phase 2 Task 4, see Figure 4-9). The primary objective of the portfolio risk assessment was to provide:

- An assessment of pertinent asset data for the purposes of risk scoring;
- Initial risk scores for assets to identify high, medium and low risk assets that can be used for prioritizing maintenance and renewal activities;
- A documented process for conducting future asset risk assessments in order to provide repeatability; and
- The methodologies for calculating asset risks in order to build the capacity of staff using the processes.

In practice, the risk assessment of the utility's asset portfolio consisted of two primary tasks:

- ***Task 1 – Asset data assessment***, which includes data compilation and assessment and development of an asset list and hierarchy
- ***Task 2 – Initial high level asset risk scoring***, development of probability and consequence of failure scores, compilation and review of risk scores and training.

These tasks are described in more detail below.

4.8 Initial Data Assessment

An initial data assessment was made in relation to DUS's water, wastewater and reclaimed water pipeline and facility assets. Data for review were provided by DUS from their Computerized Maintenance Management System (CMMS) and the Geographic Information System (GIS). For pipeline assets, the following data elements were examined:

- ***Probability of Failure,*** including CCTV/condition scoring, work orders (repair/replacement), age, size, capacity, material and pressure;
- ***Consequence of Failure,*** including traffic, soil type, pressure, length, depth, diameter, environmental-related proximity issues (waterbodies, parks, wildlife, etc.), economic-related proximity issues (commercial zones and key customers), social/ health and safety-related proximity issues (school, hospital, police, fire, etc.);
- ***Redundancy*** using network maps.

4.9 Risk Scoring Methodology

The overall approach to developing the DUS asset risk scores is illustrated in Figure 4-10. The key steps shown in this figure were:

- ***Compile and Analyze Existing Asset Data***. Available pipeline and facility asset data were compiled for the potable water, wastewater, and reclaimed water systems. Pipeline data were compiled from an existing GIS database, and facility data were compiled based on a review of available design documentation and discussions with DUS staff.
- ***Develop Asset Probability of Failure Scores***. Asset probability of failure scores were assigned to each individual pipeline segment and to the treatment process units and all network facilities across the DUS asset base. Condition scores were assigned based primarily on discussions with DUS staff and analyses of asset age and material type. For pipeline assets, a scale of 1 to 5 was used, with "1" representing Very Good

condition, and "5" representing Very Poor condition. For facility assets, a scale of 1-10 was used, with "1" representing Very Good condition, and "10" representing Very Poor condition. For the facility assets, redundancy was also examined to refine the probability of failure scores.

- ***Develop Asset Consequence of Failure Scores***. Asset consequence of failure scores were assigned to each individual pipeline segment and to the treatment process units and all network facilities across the DUS asset base. Consequence of failure scores, indicating the extent of economic, social, or environmental impact that might occur if an asset were to fail, were developed based on an analysis of facility and pipeline proximity to sensitive environmental areas, critical business or health centers, high traffic roadways, and a review of sensitive or critical customers in the DUS service area. High, medium and low consequence of failure scores were developed and translated into a 1-3 scale, with "3" representing high and "1" representing low.
- ***Calculate the Risk of Structural Asset Failure***. Risk ratings were calculated for each individual pipeline segment and for all major network facilities and the treatment process units. Asset risks were calculated as the product of the scores for probability of failure and consequence of failure. For the pipeline assets, risk scores ranged from 1 to 15; for the facilities, the risk scores ranged from 1 to 100.

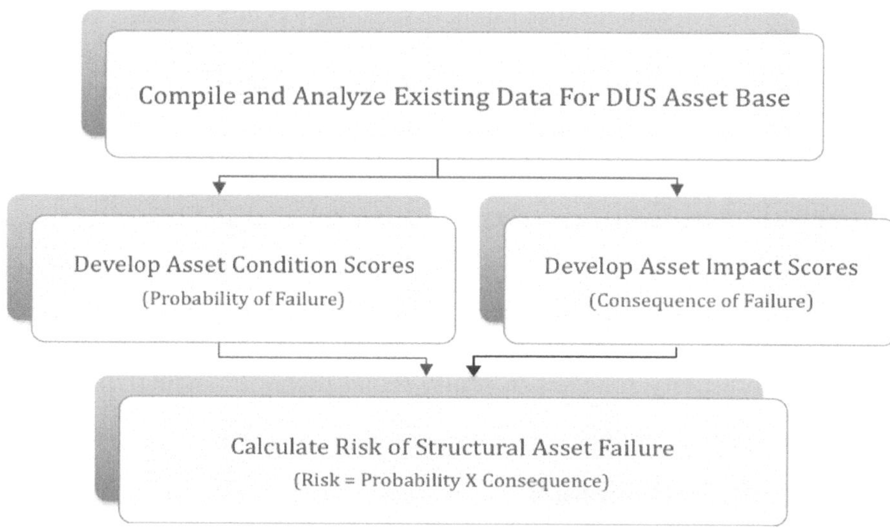

Figure 4-10. Overall Approach to Risk Scoring.

4.10 Next Steps

As a result of this initial Asset Risk Scoring effort, the following recommendations were made to DUS to enhance its wastewater, water and reclaimed water asset renewal planning and decision making capabilities:

- ***Validate High Risk Scores and Develop Mitigation Approaches:*** It was suggested that further validation of the risk scores (starting with the high risk scores) should be undertaken as part of subsequent capital improvement planning efforts and ongoing inspection and maintenance programs. Where high risk assets are confirmed, risk mitigation strategies would need to be developed, including a range of capital and non-capital improvement options.

- ♦ ***Develop Replacement Costs, and Expected Lives for Pipeline and Major Facility Assets:*** Coupled with expected life information, this information allows for development of risk-based inspection and renewal forecasts as defined in the next item.
- ♦ ***Develop Risk-Based Inspection and Renewal Forecasts:*** Long-term forecasts for the DUS assets would be developed using commercially-available decision support tools. Where available, local inspection and renewal cost data would be incorporated to generate a risk-based forecast of expected inspection and renewal costs. The decision tool would enable the timing and priorities for pipeline inspection and renewal to be aligned with DUS's preferences and risk tolerances, as well as schedule and budget constraints. These tools would also facilitate evaluation and communication of risk/cost tradeoffs with asset renewal decisions.
- ♦ ***Refine, Document, and Drive Business Processes to Enhance Scoring and Decision Making Over Time:*** Work with DUS staff to refine the asset renewal planning and data management business process adjustments that would be needed to maintain and update the asset databases with new data pertinent to risk scoring and renewal planning.
- ♦ ***Revisit and Enhance Risk Scoring Process on Ongoing Basis***: Conduct additional workshops with staff on a periodic basis to confirm risk scoring and data management processes, and refine the decision model risk matrices to meet the needs of DUS staff and decision-makers.

4.11 Research Challenges: Risk

From the information presented in this chapter, the challenge associated with assessing remaining life can be summarized thus:

1. Develop techniques for assessing the probability side of risk, including the use of condition-based and performance-based decay curves, benchmarking of decay curves to real life asset failures, and the examination of the factors that underlie the decay curves.
2. Develop protocols for risk assessment of the main assets types, such as pipes and pumps, at each stage of asset management planning using a range of techniques of varying levels of sophistication and data needs.
3. Further develop techniques for using Life Cycle Costing in combination with risk assessment, including the collection of life cycle cost data; specification of needs for a range of asset types; and dealing with uncertainty in decision-making.
4. Identify approaches for pragmatic inclusion of unrealized risk in decision-making.
5. Consider the impact of different repair and rehabilitation techniques on the probability of subsequent asset failure.
6. Improve methods to consider external costs and externalities in asset management decision-making, including methods to place a monetary value on these costs.

7. Develop techniques for assessing the consequence side of risk, including approaches to take into account the probability-side of consequence and/or appropriate ways of determining expected or maximum consequences.
8. Develop standardized protocols for assessing the importance of assets, and guidance for protocol use.
9. Use reliability approaches to determine failure modes for key asset types, and determine which of these are significant enough to warrant inclusion in further research.
10. Determine the change in failure probability that can be expected from different rehabilitation options, so that this change can be rigorously used in cost-benefit analysis of intervention options.

CHAPTER 5.0

THE ROLE OF REMAINING ASSET LIFE IN ASSET MANAGEMENT

5.1 Summary

This chapter outlines the different spatial and temporal scales of asset management, categorized in terms of the long-term company-wide view ('strategic asset management' or SAM), the medium-term sub-system view ('tactical asset management' or TAM) and the day-to-day management of assets and systems ('operations and maintenance' or O&M).

Asset remaining life concepts are applied at each of these levels of asset management. For example, in SAM, capital budgets are partly determined with respect to assets that are approaching the end of their life. Similarly, in TAM remaining life concepts are applied in prioritization and justification of capital works. Finally, various tasks such as inspection and monitoring are undertaken as part of O&M to understand remaining life.

To illustrate these different perspectives of asset management, the chapter includes case studies developed with two water authorities from outside of the U.S. – one in Australia and the other in the UK. The case studies illustrate how these companies approach each level of asset management. A brief synthesis of the commonalities and differences in approaches is also presented.

5.2 Introduction

The previous chapter considered the role that risk plays in defining and determining remaining asset life. By necessity, many of the concepts presented relate to the role remaining asset life plays in asset management. Nevertheless, an explicit consideration of what asset management is, and what role remaining asset life plays in different aspects of asset management, has not yet been presented. This chapter addresses the linkage between asset management and remaining asset life, and presents information on the following:

- The various aspects of asset management (strategic, tactical and operational) and the role remaining life plays in each;
- The approaches used by two case study partners to address the different levels of asset management for their sewer assets;
- A synthesis of these different perspectives; and
- A summary of anticipated research challenges arising from the information presented.

5.3 The Role of Remaining Life in Asset Management

A list of 'asset management' questions was presented in Chapter 1.0, which indicated the central role remaining asset life plays in asset management. In a similar vein, Vanier (2001) introduced an approach to help Authorities develop an asset management plan, expressed as six questions, and referred to as the six "whats" of asset management (see also Vanier and Rahman, 2004):

1. What assets are owned?
2. What are they worth?
3. What is the deferred maintenance? (In this context, deferred maintenance is the amount of expenditure required to bring the maintenance and repair under control, rather than being a measure of renewal backlog).
4. What condition are the assets in?
5. What is the remaining service life of the assets?
6. What should be fixed first?

While more sophisticated frameworks for asset management have subsequently been developed (e.g., the frameworks given in IPWEA, 2006 or BSI, 2008), these questions highlight the fundamental components of asset management and the importance of understanding remaining asset life to achieving its aims.

In practice, however, asset management is more complex than this series of questions suggests, and involves activities at a range of spatial and temporal resolutions, as illustrated graphically in Figure 5-1 (after Marlow and Burn, 2008). Similarly, the concept of remaining asset life varies depending on the level at which asset management is being undertaken, as described in the next sections.

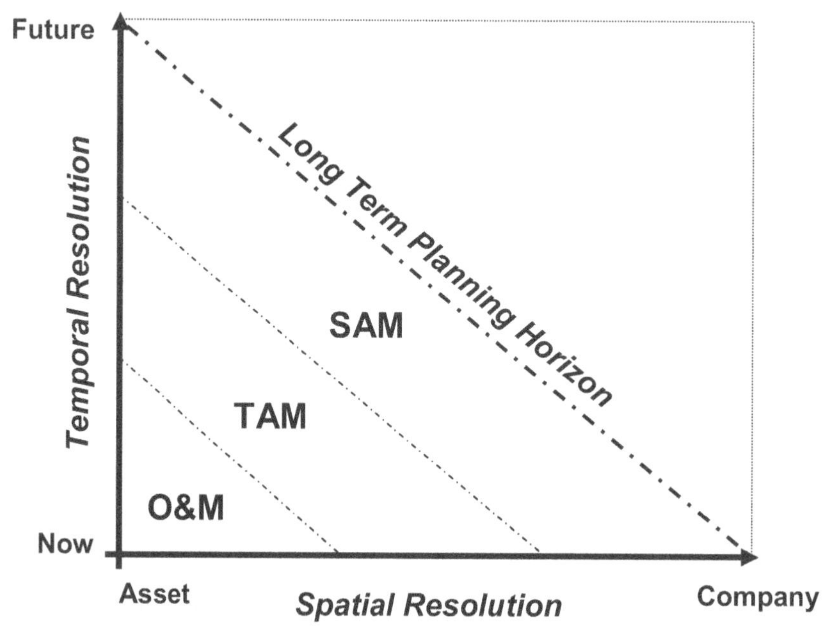

Notes: O&M: Operations and Maintenance; TAM: Tactical Asset Management; SAM: Strategic Asset Management

Figure 5-1. Different Scales of Asset Management.
(Reprinted with permission from Marlow and Burn, 2008)

5.3.1 Strategic Asset Management

The expression "Strategic Asset Management" (SAM) is often used to refer specifically to asset management practices with a long-term view. SAM is concerned with setting overall strategy and budgets using broad estimation tools, and is generally applied at a coarse resolution using spatially aggregated data and broad management units; e.g., distribution assets, transmission assets, and treatment work assets (Marlow and Burn, 2008). Within SAM, remaining asset life is considered as part of budget setting and valuation of assets.

As discussed in Appendix A, valuation of assets requires that depreciation is taken into account, which requires an assumption (age-based or use-based depreciation) or assessment (condition-based depreciation) of remaining asset life.

Determining budgets requires information on the remaining life of facilities, large individual assets or cohorts of smaller assets to be determined, along with potential intervention options (Burn et al., 2007). In practice, this requires a staged approach:

1. Metrics that reflect the current and future performance of the asset stock (including levels of service) are compared to targets.
2. Current and anticipated shortfalls between requirements and targets are identified; this shortfall defines the end of life for the assets involved.
3. Different options for addressing the gaps are considered and economic analysis undertaken to select the most appropriate intervention.

Figure 5-2 illustrates SAM in terms of this 'gap analysis' procedure. In practice, the gap analysis is undertaken to compare the capabilities of existing assets to levels of service targets and other standards, as embodied by the water authority's policies, targets and standards. The shortfall in capabilities is considered in terms of service and/or cost and can be real (associated with historical problems) or probabilistic (e.g., an unacceptable risk of asset or service failure).

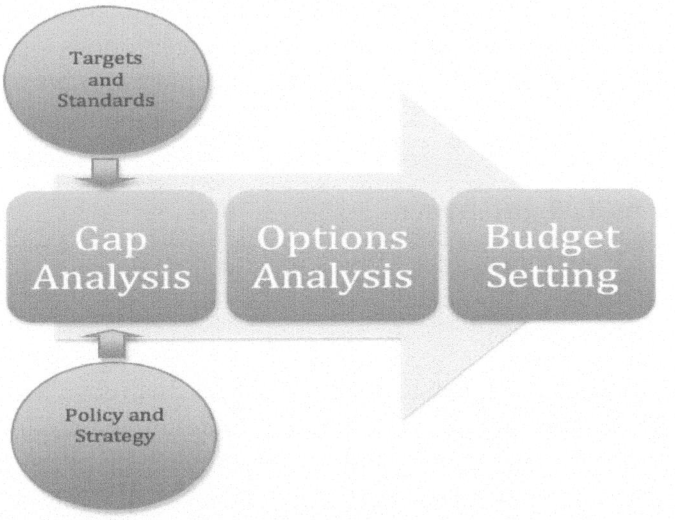

Figure 5-2. Schematic Representation of SAM.

Shortfalls can arise due to three drivers:

- ***Supply demand balance***: the capacity of the assets is not balanced against the supply side demands.

- ♦ ***Quality***: the desired service quality, especially in terms of customer or environmental metrics, is not being met. This can be associated with changes in service requirements, such as those imposed by new regulations.
- ♦ ***Aging assets***: this driver is inherently associated with the end of asset life.

Shortfalls in any of these areas can lead to assets being considered at the end of their life. Analysis can also determine when these shortfalls will occur, which defines the remaining asset life of the associated assets.

5.3.2 Tactical Asset Management

The term "Tactical Asset Management" (TAM) refers to asset management practices with a medium-term view. TAM is concerned with setting priorities and determining how, when, and where to intervene in the asset stock. TAM is applied at a finer resolution than SAM, so more detailed data are required with more specific management units; e.g., refined cohorts of assets and at the individual asset level (Marlow and Burn, 2008).

Whereas SAM is concerned with setting overall strategy and budgets, tactical asset management is concerned with delivery of capital solutions. Within TAM, activities that relate to remaining asset life thus include prioritization and justification of planned replacement work (i.e., developing business cases for proposed capital projects). Specific assets must therefore be identified that are approaching the end of their life and need to be replaced or rehabilitated in the short to medium term due to financial, service, risk or other drivers.

Prioritization of capital works may involve inspection of the asset or the application of other screening tools (e.g., Davis McDonald and Zhao, 2001; WERF, 2007; and Marlow, 2008). This prioritization has to be undertaken within the context of technical considerations, but also with a view to any political, administrative or budgetary restrictions (Vanier and Rahman, 2004). Another task undertaken within TAM is determining the most appropriate interventions for planned capital projects, including the selection of rehabilitation technology where appropriate (e.g., NGSMI, 2003).

5.3.3 Operations and Maintenance

Day-to-day asset management activities include operational risk management, capital delivery and/or maintenance, all of which require asset-specific data; e.g., for a specific pipe of interest (Marlow and Burn, 2008).

Within this aspect of asset management, activities that relate to remaining asset life include inspection and monitoring tasks, responding to failure events and service issues, making repairs, collecting data for failure events, and determining which of the failed assets need replacing. Operational and maintenance staff will also have input into decisions to modify inspection/monitoring or other risk management approaches (contingency planning, etc.) for those assets that are deemed to be approaching end of life.

5.4 Case Studies

As part of this research, case studies were developed with two Authorities to highlight the approaches taken to asset management and remaining asset life outside of the U.S. These are presented below and specific effort is made to map the approaches of the Authorities onto the three-tiered representation of asset management (SAM, TAM, and operation and maintenance). Case studies were undertaken with the following authorities:

- United Utilities, UK
- Sydney Water, Australia

The case studies draw upon information provided by each of the case study partners, supplemented through an understanding of their asset management procedures developed during face-to-face interactions with members of their asset management teams.

5.5 A UK Perspective: Sewer Management with United Utilities

United Utilities (UU) is a privatized water company operating in the northwest of England; the company's shares are listed on the London Stock Exchange. UU manages and operates water and wastewater networks that serve a population of around seven million. United Utilities also uses its core utility skills to manage other infrastructure assets, through its United Utilities Contract Solutions business.

5.5.1 Regulatory Environment

The water sector in England and Wales was privatized in 1989. The operations of the privatized companies are, however, highly regulated. In particular, the Water Services Regulation Authority (Ofwat) regulates economic activities relating to the provision of water and sewerage services. It is Ofwat's role to set minimum levels of service and determine price limits. Ofwat makes decisions independently of the UK government and works closely with other bodies including:

- The Secretary of State for Environment, Food and Rural Affairs and the Welsh Assembly Government;
- The Consumer Council for Water, which is an independent organization that represents customers' interests and deals with complaints;
- The Drinking Water Inspectorate, which sets standards for the quality of drinking water;
- The Environment Agency, which regulates and enforces water abstraction consents and quality standards in inland, estuarial and coastal waters; and
- Natural England and the Countryside Council for Wales, on environmental issues.

Asset management plans (AMPs) are produced by each water company in England and Wales every five years (this cycle is referred to as the 'AMP cycle'). These AMPs provide Ofwat (and other stakeholders) with the company's justification for desired levels of investment/maintenance (and thus the necessary customer charges). AMPs are assessed by Ofwat according to the following criteria:

1. Evidence relating to the impact of historical levels of maintenance;
2. Analysis relating to why the future is different;
3. Any scope for improvements in efficiency; and
4. Any overlap between capital maintenance and enhancement (quality, capacity, etc.) programs.

Ofwat make a price determination based on the evidence provided in AMP, along with other considerations such as scope for efficiencies.

5.5.2 Asset Management Frameworks Applied

In common with many other utilities in the UK, UU apply two national asset management frameworks of relevance to sewerage infrastructure. The first, the UKWIR Common Framework for Capital Maintenance Planning, is a nationally adopted risk-based framework for planning investment in water and wastewater assets. The second is an Industry standard approach to the management of sewerage assets; the WRc Sewer Rehabilitation Manual (SRM). These frameworks are briefly outlined in the following subsections.

The UKWIR Common Framework

The UKWIR Common Framework (UKWIR, 2002) is a framework for planning investment in water and wastewater asset stocks, and in particular the investment needed to deal with aging assets. (Investment to replace aging assets is called 'capital maintenance' in the UK, since it is maintaining the capital value of the asset stock.) The Common Framework is a risk-based approach that requires capital investment be justified on the basis of current and future probability and consequence of asset failure, with a specific focus on the impact of assets on service provision.

The key aspect of the Common Framework is that it is forward-looking and based on economic analysis of risk. The Framework specifies a number of discrete stages to be followed when assessing Capital Maintenance (and other investment) needs. The stages effectively reduce to three key steps:

1. Analysis of historical performance (service delivery given past levels of investment);
2. Analysis of future investment-needs under a range of scenarios, including the 'do nothing' option; and
3. Justification of why the future is different, which reduces to a justification for price increases based on investment needs.

The Sewer Rehabilitation Manual (SRM)

The Sewer Rehabilitation Manual (SRM; WRc, 2001) is a framework for assessing the condition and performance of sewerage networks and determining what needs to be done to address service issues identified. The SRM sets out a strategy that focuses investigations on those parts of the sewerage system where problems are most severe. The approach was developed principally to meet the sewerage rehabilitation needs of the UK water sector, but has since been applied in a number of other countries.

The basis of the SRM procedure is the systematic investigation of drainage areas. As such, the division of the utility's area into one or more drainage areas (catchments) is an essential step. A key aspect of the investigation is the structural condition assessment of the sewers, which is achieved through standard CCTV inspection and grading procedures. Selection of sewers to inspect is targeted towards those sewers where the risk of failure is highest.

Under the SRM, proactively managed sewers are defined as sewers where the potential cost of asset failure means that proactive management efforts are justified; i.e., where the cost of preventive action is much less than the costs incurred should the sewer fail. These sewers tend to be defined by asset size, construction, depth, proximity to sensitive water courses, and similar factors that influence the consequence of failure.

5.5.3 SAM, TAM, and O&M Within UU

As noted previously, asset management can be considered to be a set of activities undertaken at different scales. UU's overall asset management approach contextualized in these terms is represented schematically in Figure 5-3 and discussed in the following subsections

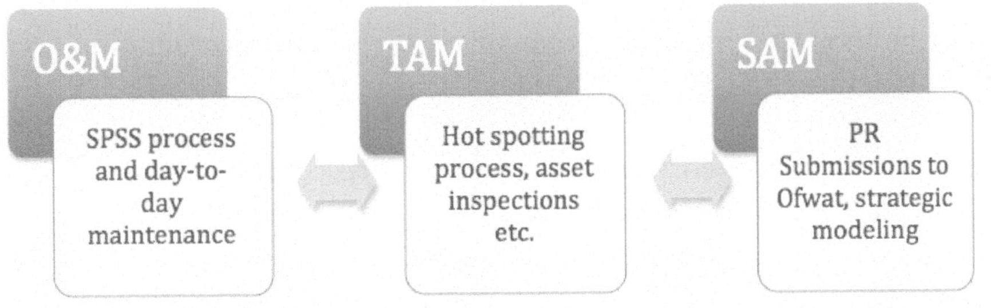

Note: O&M: Operations and Maintenance, TAM: Tactical Asset Management; SAM: Strategic Asset Management

Figure 5-3. Schematic Representation of UU's Three-Tier Asset Management Approach.

Approaches to SAM

In line with the requirements of Ofwat, UU currently analyzes investment needs over a 25-year planning horizon. Price determinations (and thus maintenance and investment levels) are, however, set over a shorter five-year period. Analysis to justify the required level of capital and operational expenditure is undertaken by UU and submitted to Ofwat in the form of an AMP.

Various modeling approaches have been used for this strategic analysis. At the time of the case study, UU was developing an in-house modeling procedure called *Pioneer* to undertake the analysis of investment needs for sewers. Pioneer models the relationship between service provided (measured through various Service Indicators) and level of investment needed to maintain that service. Service indicators (service-related KPIs) considered include: measures of flooding (internal, external), odor, pollution incidents, infiltration, blockages, collapse, noise and visual impact, and health and safety. The Pioneer process can be summarized thus:

1. For each service indicator (KPI) to be modeled, cohorts of assets are defined in terms of a range of various relevant factors; for example:
    - Asset related factors,
    - Environmental factors, and
    - Factors that reflect the demographics of the population served by the asset.
2. Analysis is then undertaken to define service-cost-time relationships for each cohort. These relationships show how service level and asset spending are related over time.

Pioneer mirrors the way in which the investment needs of UU are judged by Ofwat (i.e., in terms of cost, risk, and service provision).

The conceptual modeling process is shown schematically in Figure 5-4. The unit of analysis to be applied in Pioneer is the 'asset cohort', and cohorts are defined specifically with respect to each service indicator. With reference to Figure 5-4, Pioneer is based on the recognition that without periodic investment, service provided by a cohort of assets deteriorates over time (hence SI_{T1} represents better service than SI_{T4}). However, investment in the asset stock, such as that shown at T3, restores the ability of the cohort to deliver service.

In practice, the Pioneer approach is a staged one in which asset deterioration is predicted as a precursor for predicting changes in service levels, as outlined in Figure 5-5. (Pioneer includes separate deterioration models for collapse and blockage.) Pioneer also allows optimization based on the relative cost of investments and risk/impact of service failures, within the constraints of available resources and budgets. Ultimately, the output of the model is strategic predictions of service levels and expenditure requirements.

In summary, the concept of deterioration modeling within Pioneer is moving away from just considering asset deterioration, to explicit consideration of service deterioration. This approach still involves consideration of assets, but has a broader focus.

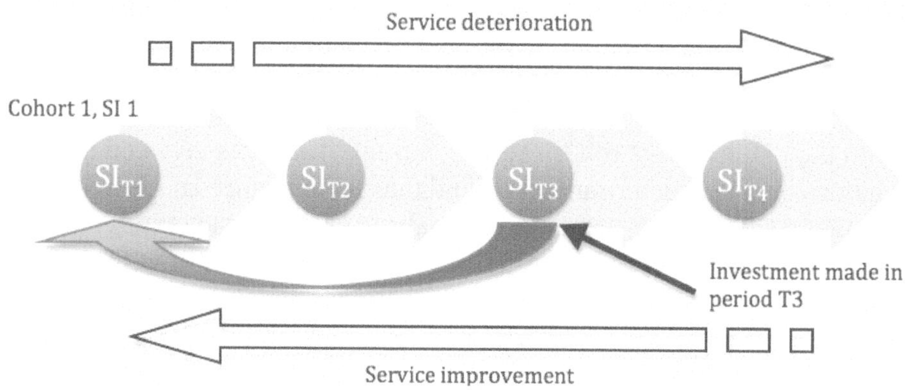

Figure 5-4. The Overall Aim of UU's Pioneer Modeling Approach.

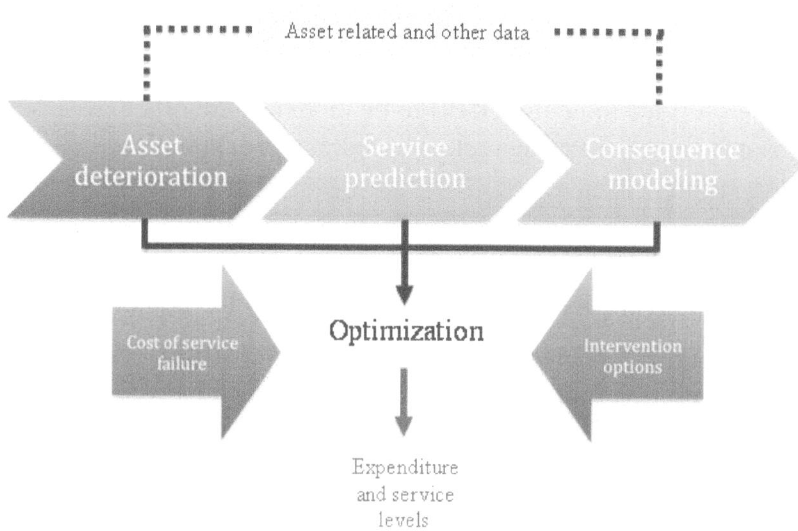

Figure 5-5. The Pioneer Approach to SAM Modeling.

Approaches to TAM

UU does not generally rehabilitate whole areas of the sewerage network. Instead, individual sewer lengths are repaired or renovated as the need arises.

For large diameter sewers, maintenance activity, including renewals, is undertaken in line with the findings of CCTV inspections. Rehabilitation techniques are selected using economic analysis.

For smaller diameter pipes, UU adopts a reactive approach and uses two targeting procedures to determine areas of the network where maintenance effort should be focused. The first of these is a tactical approach based on cluster analysis of historical blockage and collapse failures (termed 'hot spotting'). The second is based on customer complaints (termed the SPSS process using a tool called Clementine), which is as much an operational response as a tactical management procedure, so is covered in the next subsection.

The 'hot spotting' process is undertaken through analysis of blockage and collapse data (generally using data for a period spanning the previous two to five years). Cluster analysis is undertaken on this data to identify areas of the network with high levels of failures. These areas are then targeted for inspection, with defects being rectified by contractors. This rectification work is carried out within a 12-month period as a rolling program.

Approaches to Operations & Maintenance

UU's contractors undertake necessary maintenance and inspection work, including responding to service and asset failures. For proactively managed sewers, maintenance activities involve CCTV inspection and silt removal. Inspection of proactively managed sewers is undertaken routinely in accordance with the SRM approach referred to previously.

UU also uses the SPSS process (named for the software company that produced the Clementine tool) to identify areas of the network where maintenance contractors should focus effort. The SPSS process was introduced in 2006 to analyze customer complaints data from the preceding week and thereby determines problem 'polygon areas' of the network (complaints data can relate to any service issue, such as restricted toilet use, wet spots in gardens, odor issues, etc.). After the problem 'polygon area' has been identified, contractors undertake CCTV inspections on all sewer pipes within that area. Any significant defects identified are rectified. Contractors use a range of interventions, from local repair to manhole-to-manhole lining. The choice of intervention is based on UU engineer's and the contractor's knowledge of defects and their rectification, and the available repair/renovation techniques.

The rectification work is done immediately where possible, or within a matter of weeks if there are access issues to consider (e.g., road traffic management requirements). UU anticipates that the SPSS process eventually will detect problem areas as asset failures and service issues start to occur, so hot spotting analysis may not be required in the future.

5.5.4 Use of Remaining Asset Life in Asset Management

The regulatory regime in England and Wales (E&W) requires that sewers (like water mains) be maintained in perpetuity. In practice, this means that network (pipe) assets are not depreciated in the same way as discreet assets (like pumps, tanks or treatment works). Instead, an infrastructure renewal charge is calculated that reflects the level of investment needed to maintain the network of assets.

One consequence of this approach is that the concept of 'remaining life' is not explicitly used for network assets in E&W. Furthermore, it is understood that Ofwat does not accept investment planning approaches that are based solely on estimations of remaining asset life; such approaches are in any case not consistent with the requirements of the 'Common Framework'. As such, and in common with other companies in E&W, UU does not explicitly apply the concept of remaining asset life to sewers. Instead, capital investment is undertaken on the basis of risk, informed by evidence arising from CCTV inspections and/or service issues related to customer complaints and failure events.

5.6 An Australian Perspective: Sewer Management within Sydney Water

Sydney Water is a public company owned by the New South Wales (NSW, Australia) government, which underwent corporatization in 1995. It provides drinking water, recycled water, wastewater services and some stormwater services to more than four million people. Drinking water is sourced from a network of dams managed by the Sydney Catchment Authority, then treated and delivered to customers' homes and businesses by Sydney Water. This case study focuses on approaches used in the management of sewers.

5.6.1 Regulatory Environment

Since corporatization, Sydney Water has been regulated against an operating license and a customer contract. Sydney Water is regulated by IPART (the Independent Pricing and Regulatory Tribunal), the independent economic regulator for NSW. IPART oversees regulation in the electricity, gas, water, and transport industries and undertakes other tasks referred to it by the NSW Government.

IPART's role in regulating water, wastewater and stormwater services in NSW includes setting maximum prices for these services, regulating prices for the Bulk Water supplied by the Sydney Catchment Authority, administering operating licenses for Water Service Providers, including Sydney Water, and regulating the charges that State Water and the Department of Natural Resources may levy for delivering Bulk Water. Other agencies that have a role in the regulation of Sydney Water's activities include:

- The Department of Health
- The Department of Environment and Climate Change
- The Department of Planning

5.6.2 Asset Management Frameworks Applied

There are no national frameworks specific to the asset management of water or wastewater assets in Australia. However, the NSW Government Treasury Department has produced a 'Total Asset Management' framework. This framework provides guidance on strategic management of physical assets, and was introduced to achieve better planning and management of the State's physical assets, both existing and planned (NSW, 2006). The framework applies to all assets (i.e., not just water or wastewater) owned by the NSW government. Sydney Water has developed in-house asset management processes in line with the requirements of the 'Total Asset Management' framework.

The approaches adopted for sewer management were initially in line with the WRc's SRM (see above), but have subsequently evolved to a more risk-based (rather than criticality-based) procedure, which is tailored to the Australian context.

5.6.3 SAM, TAM, and O&M within Sydney Water

As noted previously, asset management can be considered to be a set of activities undertaken at different scales. Sydney Water's overall asset management approach contextualized in these terms is represented schematically in Figure 5-6 and discussed in the following subsections

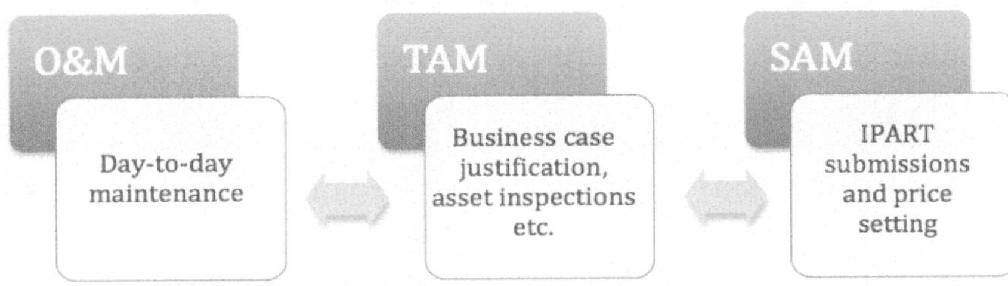

Note: O&M: Operations and Maintenance, TAM: Tactical Asset Management; SAM: Strategic Asset Management

Figure 5-6. Schematic Representation of SW's Three-Tier Asset Management Approach.

SAM

Sydney Water's approach to SAM follows the price setting cycle dictated by the regulator IPART. The price setting cycle varies in length from between two to five years, in line with the NSW's state government political cycle. The overall approach to SAM is to develop 'asset class plans' for various asset types (e.g., sewers, water mains, pumping stations, treatment works). These plans have a service and asset life cycle focus, and are used as the basis for setting prices (charged to the customer) for a specified period (two to five years). Short to medium term asset replacement needs are identified within each Asset Class Plan, but there is also a focus on how this investment will allow longer term objectives (five to 30 years) to be met.

Overall, SW is required to deliver service objectives within the price limit set by IPART. IPART engages auditors to review both the way in which assets were managed in the previous cycle, and to determine what is planned in the next cycle.

TAM

'Asset class plans' include an assessment of short- to medium-term asset replacement needs. Any capital renewal project must, however, still be approved through the annual business cycle undertaken within Sydney Water. Hence, while the 'asset class plans' do include anticipated renewal projects, what is actually delivered will depend on business priorities and other drivers. The development of business justifications for asset replacements can be considered a TAM procedure.

Operations & Maintenance

Sydney Water staff undertake necessary maintenance and inspection work, including responding to service and asset failures and making repairs. For reactively managed assets, a key issue is clearing blockages through jetting, root cutting and foaming. Structural collapse of smaller sewers must also be managed, but is a much less prevalent issue (e.g., 0.01% of sewers are subject to collapse in a year, whereas approximately 5% of sewers will suffer a blockage of some kind).

For proactively managed sewers, maintenance activities involve CCTV or walk-through inspection. Proactive assets are inspected on a 10-year cycle (maximum). The exception is large diameter concrete sewers in defined risk areas, which are inspected on a five-year cycle. Inspections can be carried out earlier when the results of the last inspection indicated that this is necessary; i.e., SW undertakes more frequent inspections as a sewer approaches the end of its life.

5.6.4 Use of Remaining Asset Life in Asset Management

Unlike in the UU case study, Sydney Water explicitly uses the concept of remaining life in their asset management procedures. In particular, the concept of service life is used in the assessment of when replacement or rehabilitation of an asset should occur. In this context, end of service life is taken to (or would) occur when the sewer fails structurally (collapses), is subject to multiple blockages or other (customer and/or environmental) service issues, or is at or past the point of optimum life cycle cost. Whether or not an asset is allowed to reach the end of its service life depends on customer service, financial and economic factors and other risk considerations. The asset management strategies adopted are summarized in Figure 5-7.

As shown in Figure 5-7, the end of life criteria used depends on whether an asset is managed proactively (referred to within Sydney Water as 'avoid failure' assets) or reactively (referred to as 'plan to repair' assets). The interpretation of Figure 5-7 is discussed in more detail below, but it is worth noting that proactive assets are replaced because of risk to service provision, whereas replacement of reactive assets is event driven (the asset fails and then a decision is made to replace it based on cost-effectiveness considerations).

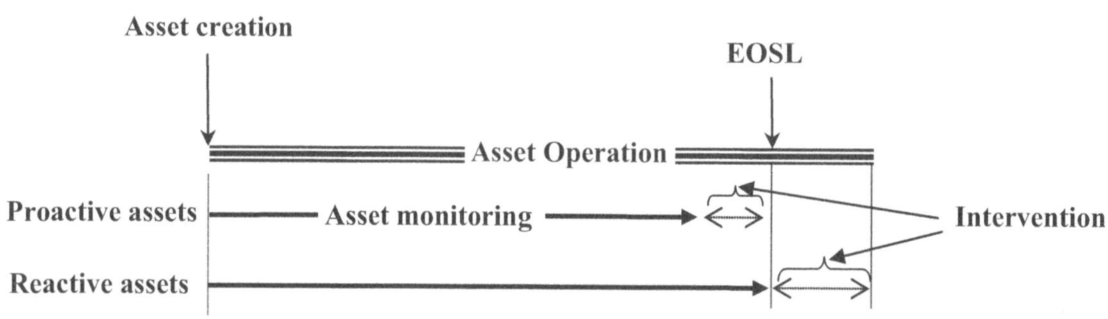

NB: **EOSL**; end of service life

Figure 5-7. Schematic Representation of Asset Life Concepts Applied by SW.

Proactively Managed Assets ('Avoid Failure' Assets)

As indicated in Figure 5-7, proactive assets are inspected, maintained and rehabilitated or replaced so that they do not deteriorate to the point where the end of service life is reached. Assets are rehabilitated or augmented when condition or performance indicates the end of service life is being approached, not after it is reached. The period between when the need for an intervention is determined and when the intervention must take place is different for different pipe materials.

Asset performance is assessed to determine if the asset is meeting its service, cost or engineering outcomes (for structural integrity and hydraulic capacity). Capacity issues can be assessed through hydraulic modeling or performance measures (KPIs relating to surcharging, flooding, etc). Insufficient capacity due to new residential or other developments can also mean there is a need to upsize a sewer.

To anticipate the end of service life due to deterioration, proactive assets are routinely inspected by CCTV or person entry (traverse) inspection techniques to identify if critical defects have progressed to the point where service may be compromised in the future. After inspection, SW assess the condition data collected and determine a remaining life according to the condition grade allocated, as shown in Table 5-1.

Table 5-1. Remaining Life According to Assessed Condition Grade.

Grade	Description	Remaining Life
1	Very good	> 25 years
2	Good	10 to 25 years
3	Fair	5 to 10 years
4	Poor	2 to 5 years
5	Very poor	1 to 2 years

The need to plan the renewal or rehabilitation of an asset is triggered when the time to end of service life is assessed as < 2 years (grade 5), except where the cost of the investment is >$1m, in which case the trigger occurs when the remaining service life is assessed as being < 2 to 5 years (grade 4 or 5). These thresholds are embedded in SW's risk-based approach for determining both the timing of investments and the type of investment required. Figure 5-8 gives more details of this approach. As illustrated, the prioritization criteria for risk mitigation such as rehabilitation or replacement, is based on the consequence of failure (cost) and years to end of service life (which is a surrogate for the probability of failure). Hence, the prioritization process is explicitly dependent on the estimate of remaining asset life determined through condition assessment (see Table 5-1 above).

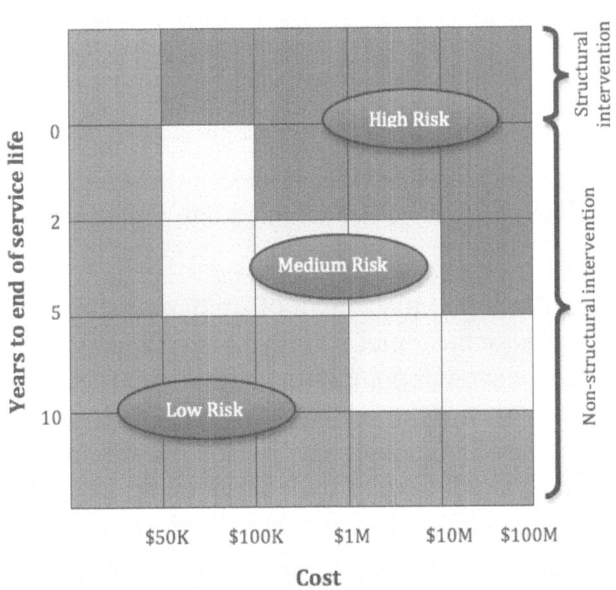

Figure 5-8. Sydney Water's Risk Prioritization Matrix.

Reactively Managed Assets ('Plan to Repair' Assets)

Reactively managed assets (which SW refers to as 'plan to repair' assets) are allowed to reach the end of their service life before they are replaced. In practice, this means that individual sewers fail (collapse, which is a relatively rare failure mode, or have multiple blockage occurrences, which is a more common mode of failure) and a decision is then made on replacement/rehabilitation based on minimum service levels and economic considerations. As with proactive assets, the need to investigate a renewal/rehabilitation option is triggered by both capacity-related and condition-related issues. For example, when:

- Sewer blockages repeat three to five years;
- Wet weather overflows resulting in flooding of homes;
- Repeat dry/wet weather overflows on private property cause outrage; and/or
- There is a high frequency of wet weather overflows reaching watercourses.

These conditions represent the end of service life for reactive assets; an economic appraisal is made when these conditions are met and a decision is made to rehabilitate or replace the asset.

5.7 Synthesis of Perspectives

One obvious difference between the approaches used by the case study partners in the UK and Australia is that UU do not explicitly use remaining life concepts in their asset management approaches. However, while there are some differences in terminology, in practice the same approach to asset management is used; i.e., a service driven, risk based, economic approach to replacement decisions.

The approaches of SW and UU to strategic asset management mirror the reporting and planning requirements of the regulator. UU base their approaches on modeling of service provision, whereas SW focuses on the development of 'asset class plans', which are built up from proposed rehabilitation projects and developed from knowledge of performance gaps, rather than modeling.

Both UU and SW use a proactive monitoring approach to determine when interventions in larger sewers are required. CCTV inspection is used to determine when deterioration has reached a trigger point and thus a capital or other intervention is required. In both cases, effort is expended to anticipate when end of service life would occur and to intervene in the most economic way.

SW has a reactive approach to replacement of small diameter sewers, which is driven by service considerations at an asset or customer level. UU have a similar approach in that their maintenance effort is driven by customer complaints or historical records of failures, but their targeting procedures identify an area of the network, which is then investigated and repaired/rehabilitated. In the case of SW, asset and service failures are used to determine when small diameter sewers have reached the end of their service life. UU have a somewhat more proactive approach in that sewers within 'polygon areas' are fixed if defects are identified during the CCTV inspection, irrespective of whether these assets were the ones creating the known service problem.

In summary, while there are some key differences in terminology, as well as modeling approaches used for prioritization and strategic planning, the fundamentals behind the approaches are consistent and include:

1. A proactive approach to large diameter or other sewers with a high potential consequence of failure, where condition monitoring is used to determine when capital interventions are needed.
2. A reactive approach to smaller diameter mains where service/asset failures and customer complaints are used to determine when capital interventions are needed.
3. A focus on making economic decisions, based on considerations of service, risk and financial and other consequences associated with the available interventions, including the do-nothing option.

5.8 Research Challenges: Asset Management

From the information presented in this chapter, the challenge associated with assessing remaining life can be summarized thus:

1. There is a need to determine suitable approaches for analysis of remaining asset life at different levels of spatial and temporal resolution to meet the needs of SAM, TAM, and operational risk management.
2. There is a need to address deterioration modeling in terms of service, and not just asset deterioration.
3. There is a need to develop protocols that consider supply demand, quality and capital maintenance issues simultaneously in the assessment of remaining asset life.
4. Standardized risk matrices for use in asset management decision-making would be useful.

CHAPTER 6.0

ASSESSING REMAINING ASSET LIFE

6.1 Summary

This chapter addresses the ways in which remaining life of wastewater assets can be assessed using data on the current state of an asset. The simplest approach is to use the age of the asset in combination with an estimate of the expected service life. For some applications, however, a more detailed assessment is justified.

Since an asset passes through a range of condition and/or performance states as it deteriorates, condition and performance assessment can be used to understand remaining asset life. In particular, acceptable asset condition states can be defined that characterize the threshold above which risk is deemed to be unacceptable. Condition assessment can then be used to determine whether the current state of an asset, expressed in terms of failure likelihood, is acceptable. Acceptability criteria for assessing degraded condition can be obtained from a number of sources including the use of expert opinion, condition grading, detection of critical defects and performance monitoring.

An engineer or maintenance practitioner with an appropriate level of knowledge and experience can make a fair assessment of whether an asset is in an acceptable state. It is also possible for the assessor to estimate remaining service life directly from information on asset condition, when the assessor has knowledge of the asset age and how the asset deteriorates over time. However, such estimates are only qualitative.

To provide more consistency, there is a need for standard guidance on what constitutes a significant defect for a range of asset types in a variety of operational contexts and/or standard methods for utilizing expert opinion. Condition and performance grading approaches have been developed to facilitate this process.

For a broad range of assets in the wastewater sector and other sectors, it is possible to specify a critical defect size that represents the limit of acceptable condition. The issue of determining acceptable asset state is then reduced to the need to assess whether such critical defects exist, or to make an assessment of the time until such defects would be expected.

Since asset condition and performance-failure can be considered as a 'cause' and 'effect' respectively, on-going assessment of asset performance with respect to acceptability thresholds provides another means of determining if an asset is in an acceptable state. Performance monitoring also provides a measure of the impact of asset deterioration, and thus helps to determine the need for maintenance and/or replacement

6.2 Introduction

The discussion presented so far has indicated that, while a range of other considerations must be taken into account, the current state of an asset is a key factor in determining remaining asset life. This chapter describes approaches for assessing the remaining life of wastewater assets using data on the current state of an asset. Information is presented on the following:

- The use of assumed asset life to determine remaining life;
- Condition as a measure of asset state for use in determining remaining asset life;
- The interpretation of condition and performance data in terms of remaining asset life;
- A case study example to show some of the difficulties in undertaking these tasks; and
- A summary of anticipated research challenges arising from the information presented.

Consideration of modeling approaches used in the prediction of remaining asset life is deferred until the next Chapter.

6.3 Assumed Asset Lives

If the age of an asset is known and its expected service life is relatively well defined, then the remaining asset life can be calculated simply by determining the difference between these two figures; that is:

Remaining asset life = (Expected service life) – (Asset current age)

As detailed in Appendix A, this approach is used in calculation of depreciation, and can be based on age or usage of the asset, as appropriate. While somewhat simplistic, this approach is based on the recognition that it is possible to specify average expected life times for a range of asset types. While the actual life of a specific asset may be higher or lower than the assumed average, over a reasonably sized population of assets, the average will still be meaningful.

This approach can be applied in valuing asset stocks since it is the overall value that is of concern; not the value of a specific asset. The fact that one asset may be over-valued and another under-valued does not matter from this perspective (Steve Whipp, UU, personal communication). While the concept of average asset life is less useful for individual assets, it is possible to modify the assumptions on the basis of asset-specific data. For example, Sydney Water uses condition assessment data to modify the remaining life for a given asset (see Appendix A).

In addition to its use in calculation of depreciation, the assumed or 'book' life of an asset can be used in decision-making with respect to the replacement of assets. For example, Hoskins et al. (1998) refer to the use of book lives to determine the age at which oil-filled circuit breakers should be replaced (a book-life of 35-40 years is given). However, as discussed in Appendix A, the actual life of an asset can often exceed the book-life. Furthermore, replacement of an asset in accordance with its book-life does not explicitly consider the other economic factors that can drive asset replacement, so such approaches do not align with the asset management principles outlined in earlier chapters.

6.4 The Role of Condition Assessment

A schematic representation of the lifecycle of an asset subject to maintenance was presented in Figure 2-1. For purposes of the current discussion, however, it is useful to consider the life cycle of an asset without the complications added by maintenance activities (see Figure 6-1). As illustrated, at the time of asset creation, there is an expected (design) life, but the actual service life of the asset can be shortened (e.g., to T_1) or increased (e.g., to T_2) by a range of factors. (See Chapter 3.0 for factors that pertain to different asset categories.)

Figure 6-1 also indicates that an asset passes through a range of condition and/or performance states as it deteriorates, and these can be labeled in qualitative terms (New, Fair, etc.). Depending on the factors that influence asset life, an asset will transition between states more or less quickly. Hence, it can be seen that the asset with life T_1 (again, as illustrated in Figure 6-1) passes through the different states more quickly than the asset with life T_2.

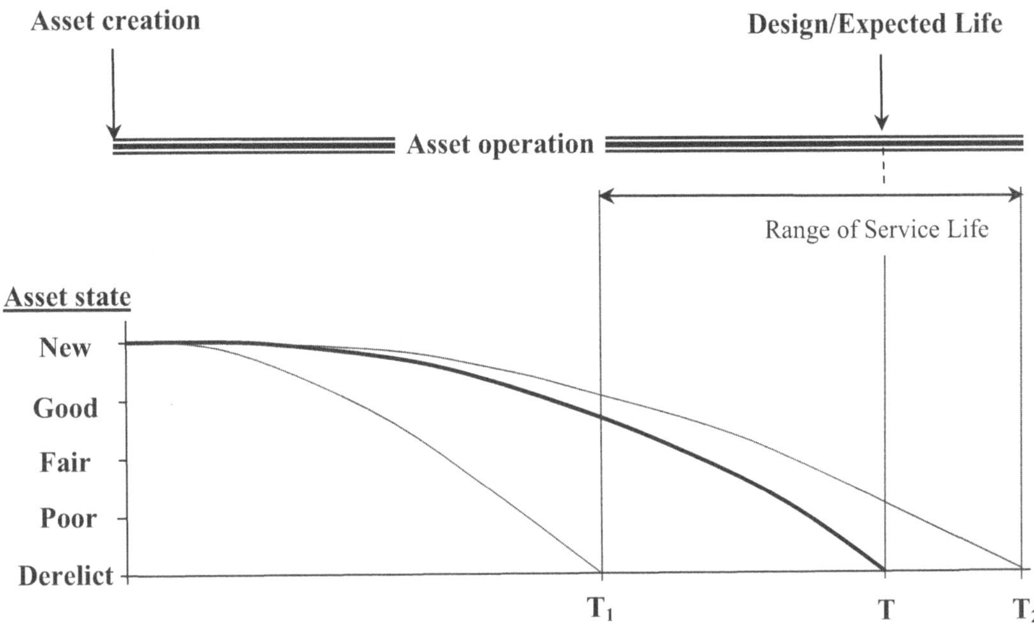

Figure 6-1. The Relationship between Condition and Asset Life.

The likelihood of asset failure tends to increase as the asset condition deteriorates. In many respects, an ideal assessment of asset state would thus be a measure of failure probability. Unfortunately, in many instances it is impractical or too costly to develop an assessment of failure probability with any reasonable degree of certainty. As an alternative, asset condition can be used as a surrogate measure for failure probability (e.g., WERF, 2007; Marlow and Burn, 2008). In this approach, acceptable asset condition states are defined, which characterize the threshold above which risk is deemed to be unacceptable (e.g., Hoskins et al., 1998). Condition assessment can then be used to determine whether the current state of an asset, implicitly expressed in terms of failure likelihood, is acceptable.

Ongoing condition monitoring (i.e., on-going assessments of asset condition and/or performance) provides an assessment of asset state over time. Ongoing monitoring also provides time series data from which the rate of deterioration can be estimated and the time to critical conditions extrapolated, which in turn gives the remaining service life (modeling approaches to achieve this are considered in more detail in the Chapter 7.0).

6.4.1 Interpretation of Condition and Performance Data

Condition assessment is the process of assessing the capacity of an asset, generally with reference to some norm (Marlow and Burn, 2008). This norm can be the 'as new' asset condition, or some specified acceptability criterion that encapsulates the boundary between acceptable and unacceptable asset states.

Acceptability criteria for assessing degraded condition can be obtained from a number of sources. For example, Esselman et al. (1998) describe approaches used in the nuclear industry, including the use of non-mandatory codes, standards and guidelines; rigorous engineering calculations; and 'accepted practice' or expert opinion. Similarly, a number of approaches are used within the wastewater sector to aid interpretation of condition and performance data, including:

1. The use of expert opinion
2. Condition grading
3. Detection of critical defects
4. Performance monitoring compared to thresholds

The Use of Expert Opinion

An engineer or maintenance practitioner with an appropriate level of knowledge and experience can make a fair assessment of whether an asset is in an acceptable state. This implicitly involves making an assessment of whether an asset is at the end of its life and requires an intervention. Newton and Vanier (2006) note that remaining service life can be estimated directly from information on asset condition, when the assessor has knowledge of the asset age and how the asset deteriorates over time. In practice, the assessor will also need to consider the asset's operating context. Such assessments may not be accurate in absolute terms, but semi-quantitative or qualitative assessments of the asset state can be made; for example, whether an asset is at or approaching the end of its life, or has a good portion of its life remaining.

The use of expert opinion to interpret inspection or condition monitoring data is a pragmatic approach, especially when data exhibit a significant level of noise such that trends are not obvious (e.g., Wang and Zhang, 2008), or when there are insufficient data to support quantitative assessments. Issues associated with the use of expert opinion include:

1. Opinion is by its nature subjective and thus depends entirely on the experience and skills of the maintenance practitioner or engineer.
2. Any assessment of remaining life will be more uncertain than an assessment of the current state of the asset.
3. Individuals are prone to personal bias. Furthermore, when interpreting defects there is a tendency for individuals to be risk averse in their interpretations and recommendations, and be influenced by recent events.

Standard guidance is needed on what constitutes a significant defect for a range of asset types in a variety of operational contexts and/or standard methods for utilizing expert opinion. Condition and performance grading approaches have been developed to facilitate this process.

Condition and Performance Grading

Grading systems have been developed that embed expert opinion on asset state and can thus be used to determine when acceptable/unacceptable thresholds of condition and performance have been reached. In conjunction with qualitative measures of failure consequence, grades can be used to prioritize interventions, including asset replacement. Two types of grade are commonly applied:

1. ***Condition grades*** are assigned through visual inspection of an asset and with reference to specified descriptions of each grade. An asset's condition grade can only be allocated reliably after visual examination of the asset. These condition grades are a measure of the level of physical deterioration with respect to the 'as new' condition.
2. ***Performance grades*** give a broad categorization of an asset's ability to function in accordance with the utility's requirements, and are assigned using operational knowledge of the asset and specified descriptions of each grade. A performance grade can only be allocated reliably with reference to detailed local operational knowledge. Grading asset performance in this way gives a measure of asset performance with respect to local (asset-level) requirements.

WERF (2007) discusses the development of grading systems in detail and presents the type of factors that are considered for a range of asset types. In summary, the design of an effective grading system involves two stages:

1. Identify asset observations that are important indicators of condition or performance (as appropriate) of the asset type in question (for example, see assessment criteria given in Table 6-1).
2. Map the asset observations to a given grading system.

Table 6-1. Condition and Performance Assessment Criteria.

Assessment type	Assessment criteria
Electrical Asset Condition	Electrically safe Level and urgency of maintenance required Visible wear and tear Condition of insulation Break downs and failure history Maintenance costs Health and safety issues Serviceability
Mechanical Asset Condition	Soundness of unit; as new? Level and urgency of maintenance required Level of wear and tear Condition of protective coatings Corrosion Break down and failure history Maintenance costs Health and safety issues Serviceability

With appropriate input from asset managers and maintenance engineers, it is possible to determine asset characteristics that reflect good or bad condition/performance for any asset type. These characteristics then form the basis of the grading system. Grading systems can be developed that are simple (Grade 1 to 5), intermediate (Grade 1 to 5 with sub grading for the worst three grades), and sophisticated (multi-faceted) ranking schemes (e.g., IPWEA, 2006), although these multi-faceted schemes can be reduced to 1 to 5 when necessary.

To illustrate the process of developing a grading system, it is informative to consider sewer grading systems commonly used in many countries, including the United States, Australia and the United Kingdom (e.g., see Rahman and Vanier, 2004). Structural condition of sewers is often assessed through closed-circuit television (CCTV) inspection (e.g., Madryas and Przbyla,

1998; WERF, 2007). A range of defects are evaluated in these inspections, including cracking, fractures, deformation, loss of fabric (including mortar loss, brick displacement, etc.), joint/connection defects and loss of level (Rahman and Vanier, 2004). A grading system must incorporate these defects in a manner that reflects the various stages of asset deterioration.

Grading approaches for sewers have been standardized, and often use a scoring procedure in which defects are given a score corresponding to the type and severity of the defect and its potential impact on asset performance and failure (Rahman and Vanier, 2004). Defects observed during the CCTV inspection are noted in a standard report that can then be run through software to score the sewer lengths and provide an overall 1 to 5 grade. This grade summarizes the condition of a sewer length, generally from manhole to manhole. The 1 to 5 grades can also be allocated directly by the inspector.

Table 6-2 shows a typical grading system that has been used in the UK, and is consistent with the grading system presented in the WRc Sewer Rehabilitation Manual (WRc, 2001) and other grading systems used in the U.S. and Australia.

Table 6-2. Ofwat PR99 Information Sewer Grading System.
(Reprinted with permission from Ofwat, 1998)

Condition Grade	Description	
	For Brick Sewers	For Other Sewers
1	No structural defects	No structural defects
2	Minor cracking or no deformation or loss of bricks and mortar loss confined to surface and line and level as built and connections satisfactory.	Circumferential cracking or moderate joint defects.
3	Deformation 0-5%, no fracture and only moderate mortar loss or displaced bricks or total mortar loss without other defects or occasional defective connections.	Deformation 0-5% and cracked or fractured or longitudinal/multiple cracking or occasional fractures or severe joint defects or minor loss of level or badly made connections.
4	Deformation 5-10% and fractured or total mortar loss or small number of missing bricks or displaced/hanging brickwork or moderate loss of level or frequent badly made connections or dropped invert.	Deformation 5-10% and cracked or fractured or broken or serious loss of level.
5	Already collapsed or deformation >10% and fractured or extensive area of missing bricks and/or displaced/hanging brickwork or missing invert.	Already collapsed or deformation >10% and cracked or fractured or broken or extensive areas of missing fabric.

The general interpretation of the condition grades can be summarized as follows:

- **Grade 1**: *New*: asset as new.
- **Grade 2**: *Good*: asset showing initial signs of deterioration, but still in good condition.
- **Grade 3**: *Fair*: asset condition generally satisfactory, except when the asset operational context implies that the deterioration could accelerate (for example, sewer prone to surcharging or in running sand).
- **Grade 4**: *Poor*: asset in poor condition; non-structural intervention can still be considered, but asset is in need of urgent attention to prevent dereliction.
- **Grade 5**: *Derelict*: asset structural integrity compromised, likely to mean structural rehabilitation or replacement is necessary.

Similar grading schemes can be developed that assess the performance of the asset in relation to current and future requirements, as discussed in WERF (2007).

In the grading scheme above, Grades 4 and 5 imply that an asset is in poor condition and requires a significant intervention. Allocation of these grades can thus be considered to indicate that an asset is either at the end of its life (Grade 5), or is approaching the end of its life (Grade 4). In general, an asset assessed to be in Grade 1 or Grade 2 would not be considered in a state that was of concern. Grade 3 assets are showing signs of deterioration. In many circumstances, assets in this Grade would be considered in an acceptable state. However, as noted earlier, other risk factors (such as running sand in the case of sewers) can imply the need for action to prevent deterioration processes from accelerating. To cater to this, for example, the WRc SRM grading approach modifies the 'Internal Structural Condition' (ICG) grade (allocated after CCTV inspection) in light of other risk factors, to give a structural performance grade (SPG). In practice, this means that a ICG 3 sewer subject to surcharging and in running sand would be re-assessed as a SPG 4 or worse sewer (see WRc, 2001 for more details).

This example indicates the need to capture factors in the grading procedure that are not explicitly related to condition. One common practice is to use a condition or performance grade as a measure of failure likelihood, and combine this with some measure of failure consequence to give a qualitative assessment of risk. For example, as illustrated in Figure 5-8, Sydney Water uses an assessment of cost consequences in conjunction with an assessment of remaining life determined through a condition grading approach to prioritize renewals. Similarly, Anderson and Civiti (2002) presented a matrix that illustrated how condition grades can be used in conjunction with a qualitative assessment of failure consequence to determine the need for rehabilitation, and thus the remaining asset life.

For the matrix presented in Figure 6-2, the assessment of remaining life (expressed in terms of time to rehabilitation) depends both on the condition grade and a qualitative consequence rating.

		CONSEQUENCE RATING		
		High Consequence	Moderate Consequence	Low Consequence
Condition Rating	5	Rehabilitate within 1 years	Rehabilitate within 3 years	Rehabilitate within 5-10 years
	4	Rehabilitate within 3 years	Rehabilitate within 5-10 years	Rehabilitate within 20 years
	3	Rehabilitate within 5-10 years	Rehabilitate within 20 years	Do Nothing
	2	Rehabilitate within 20 years	Do Nothing	Do Nothing
	1	Do Nothing	Do Nothing	Do Nothing

Figure 6-2. Pipeline Rehabilitation Decision Matrix.
(Reprinted with permission from Anderson and Civiti, 2002)

Detection of Critical Defects

For a broad range of assets in the wastewater and other sectors, it is possible to specify a critical defect size that represents the limit of acceptable condition. The issue of determining acceptable asset state is then reduced to the need to assess whether such critical defects exist, or to make an assessment of the time until such defects would be expected. In practice, the detection of a critical defect will not imply that the asset will fail immediately, but it does imply that the asset no longer conforms to a minimum acceptable condition. In this context, remaining life can be expressed in terms of the time remaining until a critical defect will be detected. Assessment of defects relative to the 'critical size' also provide a way to effectively, qualitatively and quickly "screen" equipment or structures to identify those that are approaching the end of their life (ABS, 2003).

Examples of this type of approach are referred to in the literature. For example, performance criteria related to concrete cracking of reinforced concrete structures are often expressed in terms of an acceptable crack width (ACI 318, 1999; Li et al., 2007), so crack widths above a certain limit are deemed unacceptable and require attention. Steinbuch (1999) notes that the calculation of the critical crack length is an elementary component of the safety analysis of a piping system in nuclear power plants. Similarly, Iyyer et al. (2007) discuss the concept of critical crack size as applied in aircraft life management. Datla and Pandey (2006) indicate that many utilities in North America use a minimum remaining shell thickness at the ground level as an indicator of acceptable condition for wood poles used in electric distribution systems.

These examples are given to demonstrate the breadth of assets to which the concept of a critical defect can be applied. An example of specific relevance to wastewater assets is the use of 'corrosion allowances' in the design and manufacture of engineered structures such as pipes, tanks and pressure vessels. The time when corrosion has progressed to the point where the allowance has been used up can be considered to be the dividing point between acceptable and unacceptable probability of failure. The time anticipated for this corrosion allowance to be consumed can be referred to as the "remaining life" (ABS, 2003).

The assessment procedure given in the WRc Trunk Main Management Manual (Randall-Smith et al., 1992) adopts an equivalent approach for cast iron pressure pipes, except that the critical defect is 'through wall corrosion' rather than a loss of a "corrosion allowance". The WRc assessment procedure is based on measured pit depths, using a set of simplifying assumptions, which can be summarized as:

1. Corrosion is assumed to be initiated as soon as the pipe is brought into service;
2. The internal and external corrosion rates do not change with time;
3. The deepest external corrosion pit measured is assumed to be coincident with the deepest internal pit, that is, maximum thinning of the wall, based on measured values;
4. The end of the pipe's service life is taken to occur as soon as corrosion penetrates the wall, so called through-wall corrosion.

While this approach gives some insight into remaining life across a network of assets, more detailed and accurate assessments are required when considering replacement of individual assets.

Asset Performance Monitoring

The International Infrastructure Management Manual (IPWEA, 2006) notes that asset condition and performance-failure can be considered as a 'cause' and 'effect' respectively, in that deterioration of condition is a cause of failure, and the effect of failure is poor asset performance. On-going assessment of asset performance against acceptability thresholds therefore provides another means of determining if an asset is in an acceptable state. Performance monitoring also provides a measure of the impact of asset deterioration, and thus helps to determine the need for maintenance and/or replacement (WERF, 2007).

For process equipment (M&E, ICA assets), monitoring of a range of performance parameters is a key aspect of condition monitoring (detailed consideration of this issue is beyond the scope of this report, but see Jardine et al., 2006 for a detailed review of this aspect of condition monitoring). As noted above, grading approaches have also been developed that allow the performance of an asset to be assessed.

In general, asset performance must be considered in a wider context, considering issues such as obsolescence and capacity/functional requirements. Decisions to replace an asset can thus be driven by performance issues relating to 'fitness for purpose', as well as deterioration (WERF, 2007).

6.4.2 A Case Study Example

A case study is detailed below that provides some insights into the practical problems that can be encountered when determining the remaining life of an asset based on condition and risk assessments. The case study is taken from WERF (2007), and relates to a detailed investigation into the risk and condition of an interceptor sewer crossing an environmentally sensitive area with potential for significant aesthetic and environmental impacts should the asset fail catastrophically. The interceptor sewer was 18 km long and constructed from reinforced concrete, cast in situ in sections of 30 feet (10 m), and built between 1960 and 1965. The shape and size of the sewer varies along its length, as does the earth fill above the pipe. The sections of the pipeline of specific interest to the case study were 85 inches high semi-elliptical.

Results of Initial Condition Assessments

A condition assessment of the sewer was initially undertaken using walk-through inspection techniques. Data were collected on observed defects, supplemented through photographs and notes. The distance along the asset was measured by wheeling above the waterline. The assessment indicated that the interceptor sewer was in poor condition; about 5.5 km having been subject to significant acid attack, penetrating more than 30 mm. A relatively short section of the sewer (171 m) was found to be in very poor condition; 80 mm (+/-10 mm) having been lost from the original (as-built) wall thickness of 180 mm.

This section was between two siphons (thus having limited air exchange) with a large connection discharging into it, these factors providing conditions for the generation and release of H_2S. The concrete in the section had corroded to the extent that the inner of two sets of reinforcement bars (cast within the pipe wall) was exposed in places. In addition, there was relatively little earth cover above the section, a situation that can lead to a higher live load being imposed on the sewer.

Implications of Structural Analysis

Initial structural analysis was undertaken to consider the impact of existing soil and groundwater loads, as well as traffic, on the deteriorated asset. Various loading scenarios were analyzed and the calculated safety factors compared to the requirements of applicable codes. The results of these assessments implied that there was a risk of structural failure under certain conditions, but that more information was required to refine the analysis.

Subsequent investigations into the amount of earth cover, water table depth, concrete thickness, concrete strength and soil parameters were undertaken to refine the assumptions made in the analysis. The conclusions from the refined analysis were that the sewer could safely sustain existing soil, ground water and expected traffic loads. However, the remaining wall thickness was still uncertain and results indicated that the sewer would be over stressed under certain traffic loading conditions. Significant on-going deterioration of the asset was expected to occur and increase the probability of failure over time.

Risk Mitigation and Additional Investigations

As a result of the assessment, it was determined that immediate remedial action was required using a sulfate-resistant spray-on lining system. Within the context of previous discussions then, the asset was subject to a significant level of capital investment and thus considered at the end of its life. Replacement of the asset was also considered. However, replacement was not practicable given there was insufficient capacity in the network to allow the necessary work to be readily undertaken. An expansion of the network was, however, already planned that would provide the spare capacity required to undertake the capital works.

Additional investigations were subsequently undertaken to understand the risk associated with the asset and to determine if the capital renewal could be deferred until after the additional network capacity was constructed. More refined structural analysis was undertaken using three-dimensional (3D) finite element modeling. The modeling showed that the loss of material down to the first reinforcement bar was not as significant as first thought; the pipe would lose structural integrity when there was corrosion down to the outer of the two reinforcement bars, rather than the inner. It was concluded that the level of risk associated with the asset was acceptable and that its renewal could be deferred until after the network capacity was expanded through other capital projects.

Key Issues Arising from the Case Study

This case study example illustrates two key issues associated with determining remaining asset life, namely:

- The difficulty in using defects as an indicator of failure likelihood, though end of life criteria can pragmatically be specified in this manner; and
- The need to contextualize the level of risk fully to understand the need for capital interventions, and thus the remaining asset life.

While condition assessment undertaken through visual assessment is a pragmatic technique in many instances, the presence of structural defects needs to be contextualized to understand risk fully. In this case, an exposed reinforcement bar was interpreted as a 'critical defect', being indicative of a high risk of failure. The asset was thus initially considered to be at the end of its life, but affordable replacement options were not available. Additional detailed analysis indicated the probability (and thus risk) of failure was lower than anticipated. This allowed deferral of a capital project to a time when the network could support the rehabilitation of the interceptor sewer.

6.5 Research Challenges: Assessing Remaining Life

From the information presented in this chapter, the research challenge associated with assessing remaining life can be summarized thus:

1. There is a need to demonstrate the relative benefits of using more sophisticated approaches for calculating remaining life, when compared to the use of book life and other simplified approaches.
2. There is also a need to develop guidance on what constitutes an acceptable asset state; for example, by specifying critical defects and performance thresholds that are applicable to a given asset type and operating context, and which characterize the threshold for acceptable risk.
3. There is a need to develop guidance on how best to contextualize assessments of condition in terms of risk, and to do this in a way that ensures interventions can be appropriately timed to meet asset management objectives.
4. Protocols are required for "screening" equipment or structures to identify quickly and cheaply those that are approaching the end of their life.
5. If grades are to be used, there is a need to develop consistency in grading procedures across the sector and over time, and to ensure that the grades reflect risk appropriately. There is also a need to specify protocols for how to use the grades in asset management in an appropriate manner.
6. Since much data collected in relation to condition is qualitative, subjective, or descriptive, approaches are needed that can consistently apply these data in an appropriate way.
7. Similarly, where expert opinion is used, there is a need for approaches that ensure assessments are consistent and, as far as is practicable, free of bias.
8. There is a need for a framework approach to specify what approaches to assessment are suitable, considering the cost of alternative assessment procedures and the risk inherent in the decision-making process.

CHAPTER 7.0

THE STATE OF THE ART IN MODELING REMAINING ASSET LIFE

7.1 Summary

This chapter reviews the different approaches for modeling remaining service lifetime of assets. Modeling approaches are broadly grouped into:

1) *Deterministic models – where relationships between external factors and asset performance are assumed to be certain*
2) *Statistical models – based on analysis of historical failure rate or service lifetime and other data*
3) *Physical probabilistic models – based on an understanding of the physical processes that lead to asset failure while accounting for realistic uncertainty*
4) *Soft computing or Artificial intelligence models - model structure is determined by the data and no prior relationships are assumed*

The majority of previous studies have focused on water and wastewater pipelines, but in theory, many of the techniques discussed in this chapter could be applied to other asset types.

Deterministic type models are relatively simple to develop and apply. However, they usually rely on a number of simplifying assumptions. Furthermore, deterministic models do not account for the uncertainty that is associated with asset deterioration and failure.

Statistical models attempt to capture this inherent uncertainty and use historical data describing failure rates or service lifetimes in asset cohorts. Statistical models work for assets where historical data are readily available for analysis. Bayesian analysis is a robust way of supplementing historical data with beliefs (or opinions) concerning asset failure rates or lifetimes, which are based on engineering knowledge or related observations.

Physical probabilistic models are useful in those cases where no historical data are available. These models are underpinned by a robust understanding of the degradation and failure processes that occur for an asset in service (corrosion, fracture, etc.). However, they also attempt to account for realistic uncertainty by using appropriate probability distributions for model variables. However, they can be data intensive and, in the event that insufficient data exist to describe adequately model variables, simplifying assumptions are required.

Soft-computing or artificial intelligence-based approaches are data-driven rather than model driven. Artificial Neural Networks (ANNs) predict output from input information in a manner that simulates the operation of the human central nervous system. ANNs consist of layers of nodes that provide a functional relationship between input information and predicted output. They are trained on historical data sets, which demonstrate the actual relationship between input and output information, followed by testing on independent data. While the use of ANNs to solve complex problems is increasing, they are also often considered "Black Box" solutions. Data pre-

processing, methods for determining adequate model inputs and the internal workings of ANNs are seldom considered during their application. Ideally, both expertise concerning the problem at hand, and background knowledge in ANN development should be a pre-requisite.

Fuzzy logic-based approaches are able to incorporate engineering judgment and experience in a model development process. Specifically, they are proposed to be well-suited to infrastructure problems in which data are scarce and cause-effect knowledge is imprecise. Observations and model criteria are based on vague (linguistic) terms such as 'poor', 'average' or 'good' condition. Fuzzy logic-based techniques allow the propagation of these attributes through a model, therefore yielding realistic results. As part of the artificial intelligence/soft computing modeling category, fuzzy logic-based approaches also risk being treated as "black-box" solutions. Without a thorough grounding fuzzy set theory, models can be developed that are unrealistic.

7.2 Introduction

This chapter summarizes modeling techniques that can be used to determine the remaining life of assets. The approaches reviewed are divided into deterministic models, statistical models, physical models and soft computing or artificial intelligence-based models. A summary of each technique in these categories is provided along with a description of the relevant theoretical background. Practical issues associated with the application of the technique are described together with examples of application in the field.

Those techniques that could in principle be applied to any infrastructure assets (including above ground assets) have been identified. However, it is noted that the majority of techniques covered in the literature are tailored to buried water and wastewater pipelines.

Research challenges arising from the information discussed are presented at the end of the chapter.

7.3 A Taxonomy of Modeling Techniques

Deterioration models for infrastructure systems can be grouped into the following four categories:

1. Deterministic models
2. Statistical models
3. Physical probabilistic models
4. Soft Computing or Artificial Intelligence-based models

Deterministic models are often used for phenomena where relationships between components are certain. While statistical models also assume that component relationships exist, they attempt to account for inherent uncertainties (random 'noise') that exist in model variables. Physical probabilistic models predict degradation and failure of infrastructure assets from knowledge of the actual degradation and failure mechanisms that occur in service. However, they also attempt to include realistic uncertainty in these processes by using appropriate probability distributions for model variables.

These first three categories describe the 'model-driven' approaches, in that some form of relationship is assumed between input and output data. In contrast, Artificial Intelligence-based models discriminate output patterns from a set of inputs by learning from past data, generalizing these lessons and predicting future targets (Smith, 1996). These approaches in the fourth category are therefore 'data-driven' in that the model structure is determined by the data and no prior relationships are assumed. This category of models are occasionally referred to as 'black-box' models since they appear to be only concerned with input and output data without specifying the underlying mechanism (Taylor, 1993). Table 7-1 lists the different techniques addressed in this chapter and their relevant sections.

Table 7-1. Approaches for Predicting Asset Failure Rates and Lifetimes.

Generic Approach	Specific Approach	Section
Deterministic	Empirical	7.4
	Physical	7.5
Statistical	Failure event data-based	7.6
	Service lifetime-based	7.7
	Cohort-survival	7.8
	Ordinal regression	7.9
	Markov Chain	7.10
	Bayesian	7.11
Physical probabilistic	Monte Carlo simulation	7.12
	Structural reliability theory	7.13
Soft-computing, Artificial Intelligence	Artificial Neural Networks (ANNs)	7.14
	Fuzzy Logic	7.15

7.4 Deterministic Models – Empirical Approaches

As stated above, deterministic models are often used in those instances where relationships between components are, or are assumed to be, known. Examples include physical or mechanism-based models and empirical models that relate failure rates to asset attributes. Empirical deterministic models involve fitting some form of equation to observations of asset failure. Equations can be linear or non-linear and parameters are chosen to give the best fit between predicted outcomes from the empirical equation and actual observed events. A comprehensive review of empirical deterministic models is given in the literature (Kleiner and Rajani, 2001).

7.4.1 Summary of Approach

A summary of empirical approaches to deterministic models for infrastructure deterioration are as follows:

- ♦ In theory, they could be applied to any infrastructure assets that have recorded historical failure data. However, they have been primarily applied to water pipes in the literature.
- ♦ These models can only be applied to homogeneous groups (cohorts) of assets.
- ♦ Asset related data must be known in order to segregate assets into cohorts. Historical break data are also required for each cohort.
- ♦ Typical data requirements for pipe cohorts are: pipe length, installation date, breakage history, pipe type, diameter, soil type, and failure mode.
- ♦ No commercial tools are known, but models for water pipelines are well described in the literature.

7.4.2 Theoretical Background

Empirical deterministic models involve fitting some form of equation to observations of asset failure. For example, Shamir and Howard (1979) used regression analysis to obtain a prediction of failure rates for water pipes in the U.S. The empirical equation related the number of breaks to the exponent of age as:

$$N(t) = N(t_0)\exp[A(t+g)] \tag{7-1}$$

where t is the time elapsed (from the present) in years; $N(t)$ is the number of failures per unit length per year; $N(t_0)$ is the failure rate $N(t)$ at the year of installation; g is the age of the pipe at time t and A is the coefficient of failure rate (in yr^{-1}). While Shamir and Howard (1979) provided no details on the precise location of their study or the quantity and quality of their available data, they did note that Equation (7-1) must be applied to groups of pipes (cohorts) that were homogenous with respect to the factors that influence failure.

After the initial study of Shamir and Howard (1979) a number of modifications were proposed by other researchers. For example, Walski and Pelliccia (1982) proposed that additional factors be incorporated into Equation (7-1) to capture the higher failure rates in larger diameter pipes and those that had already failed. Clark (1982) transformed Equation (7-1) to a two-phase model comprising a linear equation to predict the time elapsed to the first failure and an exponential equation to predict the number of subsequent breaks. Other modifications are reported by Marvin (1996) (a power-law as opposed to an exponential equation) and Jacobs and Karney (1994) (a linear relationship).

7.4.3 Practical Considerations

While empirical deterministic models provide a straight forward interpretation of how external factors (asset attributes) affect asset performance, they may not fit well with observed data and can only be applied to homogenous groups (cohorts) of pipes. Criteria for selection of these cohorts are rarely documented in detail in the literature. Furthermore, empirical models do not provide an improved understanding of the actual causes of degradation and eventual failure. They also do not reflect the inherently uncertain nature of asset failures, which are caused by a combination of time-dependent deterioration processes and random damage events (Morcous et al., 2002).

7.4.4 Examples of Application

Examples of the development and use of empirical deterministic models to forecast failure rates in water pipelines can be found in the comprehensive review conducted by Kleiner and Rajani (2001).

7.5 Deterministic Models – Physical Approaches

Physical deterministic models are usually developed to predict service lifetimes of individual assets, rather than failure rates of cohorts. These predictions are based on a physical understanding of the degradation mechanisms that influence the load bearing capacity of an asset and (as a result) its service lifetime. A comprehensive review of physical deterministic models is given in the literature (Rajani and Kleiner, 2001).

7.5.1 Summary of Approach

A summary of physical approaches to deterministic models for infrastructure deterioration is as follows:

- In theory, the models could be applied to any assets where the mechanism of degradation and criterion for failure is understood. However, in the literature they have primarily been applied to predict time to corrosion failure in water pipes.
- Models are normally applied to individual assets and aggregation would require identifying all 'similar' assets and repeating the analysis.
- Asset attributes and loading conditions must be known to apply models. Model variables that describe degradation (i.e., corrosion rates for metallic pipes) are usually obtained through field measurements on individual pipes. In the absence of field measured data, assumed values for model variables are required.
- No commercial tools are known, but models are well described in the literature, primarily for corrosion of metallic water pipelines.

7.5.2 Theoretical Background

Physical deterministic models traditionally have been developed to predict the time to failure for individual assets. The majority of previous research in this area has focused on the corrosion and structural failure of buried cast iron water pipes (Rajani and Kleiner, 2001).

Models vary in sophistication, addressing failure modes from simple through-wall corrosion to combined corrosion and structural failure. For example, Randall-Smith et al. (1992) proposed a linear corrosion model based on the assumption that corrosion pit depth has a constant growth rate. The remaining service life of a water main was predicted using:

$$\rho = \left(\frac{t}{P_e + P_i} \delta \right) - t \qquad (7\text{-}2)$$

where ρ is the remaining service life; t is the age of the water main; P_e is the external pit depth; P_i is the internal pit depth; and δ is the thickness of the original pipe wall. As outlined by Rajani and Kleiner (2001), this simple model was originally developed as a screening tool to identify potential problems rather than to provide a means of predicting failure. It was argued that since all the calculations were based on the most pessimistic assumptions, a result indicating remaining life that exceeded the planning horizon meant that failures from corrosion pitting were unlikely to occur within the planning period. Conversely, if the calculated remaining service life was shorter than (or close to) the planning period, more rigorous analysis would be required.

One of the shortcomings of the simplified approach of Randall-Smith et al. (1992) is the assumption of linear corrosion. Studies in the literature indicate that for a cast iron pipe in aggressive soils, external surface corrosion is non-linear. Initially, corrosion rates are high, but then slow as the outer surface of the pipe becomes protected by corrosion product. To address this non-linearity, Doleac et al. (1980) used a Rossum's (1969) power function to predict the remaining wall thickness in a cast iron water pipe subjected to external surface corrosion.

$$\delta = K_n K_a (10 - pH)^n \rho^{-n} A^a \qquad (7\text{-}3)$$

In Equation (7.3), δ is the average corrosion pit depth; a, K_n and K_a are the empirical constants derived from field or lab tests; A^a is the pipe surface area exposed to corrosion; pH is the surrounding soil pH; ρ is the soil resistivity; n is the soil aeration constant and t is the time in years. Having calculated remaining average wall thickness from Equation (7-3), Doleac et al. (1980) calculated the hoop stress in the pipe wall using Barlow's equation:

$$\sigma = \frac{pD}{2t} \tag{7-4}$$

where p is the operating pressure of the pipe; D is the outside diameter, s is the hoop stress and t is the pipe wall thickness. The criterion for pipe failure was defined simply as a "reduction in pipe wall thickness to the point where a pressure surge in the pipe, equal to 50% of the working pressure, would raise σ to the material's elastic limit." Equations (7-3) and (7-4) could then be combined to calculate the time when this failure criterion was met.

While Equations (7-3) and (7-4) may provide a criterion for combined corrosion and mechanical failure, the assumption that failure occurs under internal pressure only is questionable. To remove this limitation, Rajani et al. (1996) developed a more sophisticated model to quantify the stresses in the pipe wall that arise from the interaction between the pipe and surrounding soil under pressure and temperature change. Further improvements were also made to quantify the pipe wall stresses due to the action of frost heave on buried pipes (Rajani and Zhan, 1996) and a new failure criterion was proposed in terms of the fracture toughness of un-corroded cast iron (Rajani and Kleiner, 2001). These refinements have led to a more realistic failure model that more accurately captures the corrosion and brittle fracture processes that occur for cast iron pipes in service.

7.5.3 Practical Considerations

While physical deterministic failure models may provide insights into degradation and failure mechanisms, there is a trade-off between model complexity and usability. For example, while those simpler models (Randal-Smith et al., 1996) may be useful screening tools, they also include key simplifying assumptions that cannot be validated. An example is the assumption of linear corrosion in cast iron pipelines. At the other extreme, while the more sophisticated physical models (Rajani and Kleiner, 2001) accurately capture degradation and failure processes, their input variables may be difficult to obtain.

Regardless of their complexity, it is noted that physical deterministic models can only be applied to individual assets and do not provide estimates of failure rates for asset cohorts.

7.5.4 Examples of Application

The comprehensive review conducted by Rajani and Kleiner (2001) describes the development and use of physical deterministic models to predict service lifetimes in water pipelines.

7.6 Statistical Models – Failure Event Data Approaches

Statistical models can be applied to forecast future failure rates of assets, provided sufficiently large volumes of historical failure data are available. In general, the occurrence of failures over time is treated as a stochastic process, and is represented by an appropriate probability distribution, such as variations on the Poisson process (Jarrett et al., 2001). The underlying stochastic process model is then used to estimate expected failure rates, with model parameters set up to describe the influence of variables such as material, asset size/geometry and surrounding environmental conditions. The model is calibrated using maximum likelihood fitting methods to provide the best match between model predictions and recorded failure data. Goodness of fit between model forecasts and actual observations is then demonstrated by comparison with a blind data set that was not part of the calibration process.

The majority of models of this type have been tailored to water pipeline assets. Water utilities typically collect information on most water main failures that occur. Basic information collected by most utilities consists of pipe identifier, its installation date, pipe diameter, pipe length, mode of failure, etc. Other relevant information—such as the surrounding soil type, number and type of service connections—are not always available. Failure patterns in water mains can be affected by many factors. Pipe material and size, soil type, installation practices, etc. are considered static factors because they do not vary with time. Climatic conditions (temperatures, precipitation) as well as operational practices (CP, pressure regimes) are considered dynamic, or time-dependent, factors because they change (or can change) over time.

7.6.1 Summary

A summary of the failure event data approaches to statistical models for infrastructure deterioration is as follows:

- In theory, these models could be applied to *any* infrastructure assets that have recorded historical failure data. However, they have primarily been applied to forecast failure rates in water pipes, and to the analysis of sewer blockages.
- These models can be readily applied to asset cohorts, but further work is required to predict failures in individual assets using this approach.
- Typical data requirements for a pipe cohort are: pipe length, installation date, breakage history, as pipe type, diameter, soil type, failure mode, etc.
- Commercially available tools are D-WARP (produced by National Research Council Canada) and PARMS-Planning (produced by CSIRO Australia).

7.6.2 Theoretical Background

A number of previous studies have used historical data to forecast expected failure rates in homogeneous groups (or cohorts) of pipes. In these studies, a pipe cohort is defined as a population comprising water mains of a certain material, size, vintage, etc. Shamir and Howard (1979) were the first to suggest that an exponential function could represent the increase in water main failure rates with pipe age: they proposed a two-parameter (single variate – age) exponential function, which was subsequently used by others (e.g., Walski and Pelliccia, 1982; Clark et al., 1982; Kleiner et al., 1998; Kleiner and Rajani, 1999).

More recently, Kleiner and Rajani (2000, 2002) proposed a generalized form of the exponential model that included multiple covariates, which can be time-dependent:

$$N(\underline{x}_t) = N(\underline{\overline{x}}_{t_0})e^{a \cdot \underline{x}_t} \tag{7-5}$$

where t = time elapsed from year of reference t_0, x_t = row vector of time-dependent covariates prevailing at time t, $N(\underline{x}_t)$ = number of failures (or number of failures per unit length of pipe) at time t (resulting from $\underline{x}_t$), $\underline{a}$ = column vector of parameters corresponding to covariates $\underline{x}$, and $\underline{\bar{x}}_{t_0}$ = vector of baseline $\underline{x}$ values at year of reference t_0. This model is based on the assumption that water mains experience 'background aging', which is represented by a gradual (assumed exponential) increase in failure frequency. The term "aging" as used here has the connotation of deterioration (wear and tear) and not that of chronological age. The term "background" refers to the notion that the deterioration is due to external (soil type and its corrosivity) and/or internal environments and is not influenced or accounted by any of the time-dependent factors. The model assumes that the time-dependent factors, in addition to the background aging, influence the observed failure frequency. Mathematically, these factors, which may include climate, cathodic protection, etc., influence modeled failure frequency in the exponent.

In contrast to the exponential type of model for forecasting failure rates in cohorts, Jarrett et al. (2001) proposed an alternative method to forecast failure rates in individual buried water pipelines. Their modeling approach was based on previous research by Constantine and Darroch (1993). The approach assumes that failure rates are influenced by attributes such as pipe diameter, soil type, and pipe age, and that the effects of these factors are likely to be multiplicative. For example, (all other things being equal) an inferior set of pipes might have a failure rate, say, 20% higher over its lifetime than another set of pipes. If this is so, then it is appropriate to suppose that the number of failures follows a Poisson distribution with a mean that is multiplicative in form. Such a model assumes that, for the ith pipe type, jth installation year era and kth repair year, the number of failures is y_{ijk}, which follows a Poisson distribution with mean μ_{ijk}. As described by Jarrett et al. (2001) μ_{ijk} *is* proportional to the length laid for that material in that year, with multiplicative terms for pipe type (i), installation year (j), repair year (k) and pipe age (k - j), subject to ($k - j$) being non-negative. The logarithm of μ_{ijk} is given by:

$$\log(\mu_{ijk}) = \log(L_{ij}^{\theta}) + \alpha_{ij} + \lambda_{i,k-j} \tag{7-6}$$

where θ is a length exponent, which is used in the case that failure rates may not be proportional to pipeline length; α_{ij} represents the effects of pipe and surrounding soil attributes and $\lambda_{i,k-j}$ represents the contribution to failures from pipe age. Equation (7-6) represents a "generalized linear model" (GLM) in which the number of failures in each pipe is assumed to follow a Poisson distribution with the mean described by the above equation. The "linear" part of the description follows from the fact that the mean, on a log scale, resembles a linear model. The additional assumption in the GLM is that age effects can be modeled by the cumulative hazard function $H(t) = A\,t^{\beta}$, where β represents the rate at which a pipe deteriorates with age; $\beta = 1$ represents the case where the failure rate is constant regardless of age. The number of failures in a one-year period from age t to age (t + 1) is then given by a Poisson distribution with mean value of:

$$H(t+1) - H(t) = A\{(t+1)^{\beta} - t^{\beta}\} \tag{7-7}$$

It can then be shown that the expected number of failures in a pipe is given by:

$$\mu_{ijk} = L_{ij}^{\theta} \exp(\alpha_{ij})\{(t_{ijk}+1)^{\beta_i} - t^{\beta_i}\} \tag{7-8}$$

where t_{ijk} is the age of the ith pipe with an installation year j and a repair year k.

Jarrett et al. (2001) demonstrate how the α-parameters in Equation (7-8) can be set up to reflect the influence of pipe material, pipe size, operating pressure and surrounding soil environment. As with cohort-based approach of Kleiner and Rajani (2002), the parameters in Equation (7-8) are tuned to give the best match between the observed and predicted failure rates across the cohort under investigation.

While the approaches described above can be used to forecast expected failure rates in pipe cohorts, Jarrett et al. (2001) noted that their ability to predict failures in individual assets was limited. They observed that the variation between the expected failures in individual assets determined from the Generalized Linear Model (GLM) and the actual failures experienced by individual assets was much greater than that expected from a Poisson distribution. While the model captured the number of assets that had experienced one failure, it did not capture the number of assets that had multiple failures. To solve this problem, the researchers introduced a random component of "extra-variation" in failures that varies from one asset to another. This essentially led to a Negative Binomial distribution with a percentage of extra-variation that was obtained by matching observed failures to actual data. For example, the data in Table 7-2 show the agreement between observed (n) and predicted (E) failures in individual assets with 15% extra variation.

Table 7-2. Comparison of Poisson and Negative Binomial.
(Reprinted with permission from Jarrett et al., 2001)

Interval			Observed				Poisson				Negative Binomial			
up to:	Number	Mean	n(0)	n(1)	n(2)	n(3)	E(0)	E(1)	E(2)	E(3)	E(0)	E(1)	E(2)	E(3)
0.005	16083	0.0021	16052	29	2	0	16049.3	33.7	0.0	0.0	16051.9	28.7	2.2	0.2
0.01	11005	0.0056	10948	54	2	1	10943.5	61.3	0.2	0.0	10948.4	52.1	4.0	0.4
0.02	12394	0.0136	12238	144	11	1	12226.6	166.3	1.1	0.0	12239.7	141.5	11.4	1.2
0.03	6911	0.0256	6750	145	16	0	6736.3	172.4	2.2	0.0	6750.0	146.9	12.6	1.4
0.04	4661	0.0345	4512	138	10	1	4502.9	155.4	2.7	0.0	4515.2	132.4	11.9	1.3
0.05	3872	0.0478	3707	148	14	3	3691.3	176.4	4.2	0.1	3705.2	150.5	14.3	1.6
0.075	5736	0.0600	5425	282	25	4	5402.0	324.1	9.7	0.2	5427.6	276.8	27.8	3.3
0.1	2367	0.0841	2188	161	16	2	2176.1	183.0	7.7	0.2	2190.6	156.6	17.3	2.1
0.15	1627	0.1297	1448	151	24	4	1429.1	185.4	12.0	0.5	1443.8	159.2	20.7	2.8
0.2	318	0.2013	265	42	11	0	260.0	52.3	5.3	0.4	264.2	45.2	7.3	1.1
0.25	75	0.2133	63	9	2	1	60.6	12.9	1.4	0.1	61.6	11.2	1.9	0.3
0.3	25	0.2	21	3	1	0	20.5	4.1	0.4	0.0	20.8	3.5	0.6	0.1
0.35	10	0.4	7	2	1	0	6.7	2.7	0.5	0.1	6.9	2.4	0.6	0.1
0.4	2	0	2	0	0	0	NA	NA	NA	NA	2.0	0.0	0.0	0.0
0.45	3	0	3	0	0	0	NA	NA	NA	NA	3.0	0.0	0.0	0.0
0.5	1	2	0	0	1	0	0.1	0.3	0.3	0.2	0.2	0.3	0.2	0.2
0.6	4	0	4	0	0	0	NA	NA	NA	NA	4.0	0.0	0.0	0.0
Total	65094		63633	1308	136	17	63505	1530	47.7	1.8	63635.1	1307	133	16.1

The first row of Table 7-2 shows that there were 16083 assets for which the GLM predicted between 0 and 0.005 failures. The first set of 4 columns shows that only 31 of these experienced any failures in the period, 29 assets with 1 failure [column n(1)] and 2 assets with 2 failures [column n(2)]. The next set of columns shows the number of these assets expected to have 0, 1, 2, 3 or more failures [columns E(0), E(1)... etc.] if the assets had the same mean and a Poisson distribution, while the last set of columns shows what the Negative Binomial predicts with 15% "extra-variation". Across all the rows, it can be seen that the Poisson model under-predicts the number of assets with two failures and three or more failures, while the Negative Binominal gives a much better prediction.

As described by Jarrett et al. (2001), the Negative Binomial model leads to a different set of estimates referred to as "predicted failures", in contrast to the "expected failures" provided by the Poisson model. The predicted failures for any year are obtained by (1) calculating the expected number of failures in an asset for that year and (2) multiplying this by a factor that depends on the asset. The factor that is used in the case where the extra-variation is 15% is:

$$0.15\, y/\mu + 0.85 \qquad (7\text{-}9)$$

where y is the number of failures observed in the asset during the period of observation and μ is the expected number of failures over the same period, calculated using the Poisson model. Thus, if the asset has had no failures, the predicted number of failures is 15% less than the expected number; while if the actual number is equal to the expected number, the prediction is unchanged.

The extra variation of 15% in Equation (7-9) relates to the specific case that was addressed by Jarrett et al. (2001). A separate analysis would be required to determine the extra variation for other cases. However, details of precisely how this analysis would be performed are not well-documented.

7.6.3 Practical Considerations

While statistical-based models can be applied to assets with sufficient historical failure or condition data, their applicability is limited when considering newer assets or assets located in areas where relatively severe consequences would be incurred upon failure. In these cases, insufficient historical data often exists for confident forecasts of future performance.

The partitioning of data into homogeneous populations or "groups" warrants careful attention because two conflicting objectives are involved. On one hand the groups have to be small enough to be uniform but at the same time, the groups have to be large enough to provide results that are statistically significant. Pipe characteristics that are most likely to affect water main failure patterns significantly need to be identified, and used as grouping criteria. Ideally, candidate characteristics for grouping criteria include pipe material, vintage, and diameter, soil type, operating pressures (if static), road type, road surface condition, material and density of service connections. The typical utility, however, will rarely have all these data available in a useable form.

The extension to the cohort-type models to predict failures in individual assets have clear advantages, but to date details have not been documented in the literature

7.6.4 Examples of Application

The time-dependent multi-covariate failure prediction approach was applied by Kleiner and Rajani (2004) to quantify the effectiveness of hot spot and retrofit (galvanic only) cathodic protection to reduce failure rates in metallic water mains. Subsequently Rajani and Kleiner (2007) used this approach to demonstrate the analyses of field data and show how they can be used effectively to plan the renewal of municipal water networks.

The use of the Generalized Linear Model to forecast both expected failure rates and failures in individual assets has been reported by Constantine and Darroch (1993),Veevers and Hussain (1999) and Jarrett et al. (2001, 2002)

7.7 Statistical Models – Service Lifetime Approaches

In addition to models that forecast failure rates based on observed failure events in assets, statistical techniques are also available to analyze recorded service lifetime data. In these techniques, service lifetime is usually defined as the time from asset installation to first failure. In general, these techniques involve fitting the variation in observed service lifetime data within a cohort of assets to an appropriate probability distribution. If a suitable distribution can be chosen and goodness of fit demonstrated, it can then be used to estimate expected remaining lifetime for a representative asset in the cohort and to estimate expected average failure rates across the asset cohort.

7.7.1 Summary

A summary of service lifetime approaches to statistical models for infrastructure deterioration is as follows:

- In theory, this approach could be applied to <u>any</u> infrastructure assets where recorded service lifetime data are available. The majority of examples in the literature apply to machinery and components.
- This approach can only be applied to homogeneous groups (cohorts) of assets.
- Asset attribute data must be known in order to segregate assets into cohorts.
- No commercial tools are known, but models for machinery and components are well described in the literature.

7.7.2 Theoretical Background

If recorded service lifetime data are available for a particular asset cohort, then it remains to represent this data using an appropriate probability distribution.

As discussed by Herz (1996), various mathematical distributions can be considered for modeling probabilistic lifetimes in infrastructure components and some are more applicable than others. For example, the range of application of the Normal distribution ($-\infty$ to $+\infty$) permits a negative lifespan, which is unrealistic. Although the Logarithmic Normal distribution is skewed towards the positive range of lifetimes, the residual life expectancy calculated from this distribution can increase with age, which is also unrealistic (Herz, 1996). The exponential distribution could also be applied, but as described by Crowder et al. (1991), the corresponding failure rate calculated from this distribution is independent of age, which can often contradict the observed pattern of asset failures in service.

While the distributions discussed above may be rejected due to unrealistic outcomes, the Weibull distribution is suitable for modeling pipe lifetime data (Herz, 1996). The two-parameter Weibull cumulative distribution for service lifetime is given by:

$$F(t) = 1 - \exp\left\{-(t/\alpha)^{\eta}\right\} \tag{7-10}$$

This corresponds to the probability that asset failure will occur in the period up to a specified time t. The Weibull scale and shape parameters are α and η, respectively, and are obtained by fitting Equation (7-10) to the empirical cumulative probability distribution for asset lifetime generated from the Monte Carlo simulation. An alternative lifetime distribution is the so-called 'Herz distribution', which was developed to model uncertainty in water pipeline service lifetime (Herz, 1996). The Herz cumulative probability distribution for service lifetime is written as:

$$F(t) = 1 - \frac{a+1}{a + \exp[b(t-c)]} \qquad (7\text{-}11)$$

where *a*, *b* and *c* are constants. The parameters *a*, *b* and *c* in Equation (7-11) are chosen to provide the best match between empirical and proposed distributions. The Herz distribution has the advantage of a failure rate that increases with age approaching a boundary value (Herz, 1996). This trend is frequently observed in historical failure rate data for water pipelines. The most appropriate probability distribution for asset lifetime data varies and statistical test procedures for assessing goodness of fit are documented in the literature (D'Agostino and Stephens, 1986). Having identified an appropriate probability distribution to represent the uncertainty in service lifetime, values of expected lifetime can be estimated together with upper (optimistic) and lower (pessimistic) limits corresponding to the standard deviation of the distribution.

Having chosen the most appropriate probability distribution to model lifetime data, a hazard (or failure rate) function can then be defined as:

$$h(t) = \frac{dF(t)/dt}{1 - F(t)} \qquad (7\text{-}12)$$

where $dF(t)/dt$ is the derivative of the cumulative lifetime probability distribution with respect to time. $h(t)$ represents the percentage of an asset network that is expected to have failed at a particular time and can be used to calculate the expected average failure rate.

7.7.3 Practical Considerations

In contrast to those approaches that analyze failure event data for individual assets (section 7.6.2), service lifetime approaches consider only the time to first failure. As such, they cannot be considered as full-fledged failure rate prediction models.

Similar to failure event data approaches, service lifetime models can only be applied to asset cohorts. The criteria used to segregate assets into cohorts must reflect their influence on asset failure.

7.7.4 Examples of Application

A common use for lifetime distributions is the area of reliability centered maintenance (e.g., ABS, 2004). These applications do not refer to particular infrastructure assets and propose that failure rate for any asset can be expressed in terms of the time elapsed before the first failure occurs. A common probability distribution that is used to model equipment failures is the Weibull distribution (Equation 7-10). Following the analysis outlined above, the probability distribution is then used to determine a mean time to failure (MTTF). MTTF is the average life to failure for the equipment failure mode and is defined as the point at which the areas under the probability distribution curve are equal above and below the point. MTTF data are helpful in determining appropriate maintenance strategies. For example, if the appropriate maintenance strategy is to re-build an equipment item, then the MTTF data can be used to help set the rebuilding task interval.

7.8 Statistical Models – Cohort Survival Approaches

The process of sewer deterioration can be described by a cohort survival model. In this model, cohorts are sewers of the same period of construction sharing some other features (such as material, diameter, bedding and subsoil characteristics) which are thought to influence their service life. Within their life-span, they pass through different categories of condition, from best to worst. With some probability, they survive a number of years within a category of condition. These survival curves are, in practice, transition curves into worse categories of condition. They can be determined from inspection data and then used to forecast the number of years it takes until a specific type of sewer will enter a critical category of condition.

7.8.1 Summary

A summary of the cohort survival approach to statistical models for infrastructure deterioration is as follows:

- In theory, this approach could be applied to any infrastructure assets that have recorded data describing the asset condition. However, it has been primarily applied to sewer pipes in the literature.
- Condition grade information is required (e.g., 1 = Good to 5 = Poor).
- The models can only be applied to homogeneous groups (cohorts) of assets.
- Asset attribute data must be known in order to segregate assets into cohorts. Historical condition grade data is also required for each cohort.
- Other than KANEW, no commercial tools are known, but models for sewer pipelines are well described in the literature.

7.8.2 Theoretical Background

A cohort survival model for infrastructure deterioration was first formulated at Karlsruhe University by Herz and Hochstrate (Herz, 2002, 1996; Hochstrate, 2000). In this model a cohort consists of a group of similar sewers with the same year (or period) of construction and which share features such as material type, diameter, storm/wastewater sewer, slope and other characteristics considered to influence their service lifetime.

Within their life span, cohorts pass through different condition classes, from best to worst. The model requires condition monitoring of pipes within each cohort. Classified inspection data is used to determine the survival curves, which give the transition probabilities for each cohort as they deteriorate towards the worst condition class. These survival curves are then used to predict the number of years it is likely to take until a sewer will enter a critical condition.

Two software packages emerged from the model (Herz, 1996), one for drinking water networks (KANEW, 2003) and one for sewers (Hochstrate, 2000). Both programs use a special distribution function, the so-called Herz distribution, which appears to be appropriate for modeling the deterioration of long-lived infrastructure elements. The failure and transition rate is mathematically linked with the probability density function of the service life, the survival and transition functions. The residual life expectancies and expected time in each condition class can be obtained from these functions. This information can then be used to plan selective inspection for sewers most likely to be approaching a critical condition.

The cohort model was the subject of detailed research by Baur and Herz (2002) and investigated in a Norwegian case study (Hörold, 1998). The model has been widely applied in Germany, where it was adopted to form the guidelines for cost efficient sewer rehabilitation in Baden-Württemberg, one of the sixteen German Länder (MUV-BW, 2000).

As a cohort deteriorates, there is some probability for the number of years it takes for it to move through successive condition classes. As noted above, the deterioration process can be modeled by the so-called Herz distribution (Herz, 1996), where after some time of resistance, the failure and transition rates increase exponentially up to the median age, t_{50}, then turn into a digressive curve approaching a finite maximum value. The transition functions are calibrated by curve fitting using the least square method. The transition functions *R(t)* can be written as follows:

$$R(t) = (A+1)/\left(A + e^{B(t-C)}\right) \tag{7-13}$$

where *R(t)* is the percentage of pipes that will not have moved into an inferior condition class at age, *t*; *A* is the aging parameter; *B* is the transition frequency (1/years); *C* is the resistance time (years).

The median age is derived as $t_{50} = \boldsymbol{C} + \boldsymbol{B}^{-1}\ln(\boldsymbol{A}+2)$. At this age, 50% of sewers have moved into the next condition class. As the cohort approaches finite maximum value the most resistant infrastructure elements show no increase in failure rate and so can be said to get older without actually deteriorating.

7.8.3 Practical Considerations

A disadvantage of this approach is the use of cohorts as an analysis unit, which requires a substantial amount of data to determine a 'homogenous' cohort or group (Kleiner et al., 2007). Since these data often are not available, assumptions are required and the predictive results of these models could be imprecise.

7.8.4 Examples of Application

Five condition classes are used to describe the deterioration of a cohort from excellent condition, (condition class five, cc5) to requiring immediate rehabilitation (condition class one, cc1). Transition functions calibrated using classified CCTV inspection data are shown in Figures 7-1 and 7-2. In Figure 7-1, the transition function between condition class four (cc4) and condition class three (cc3) is shown; i.e. t_{50} is approximately 60 years.

Inspection of sewers reveals whether they have deteriorated faster or slower than the average. The aging speed of an inspected sewer is dependent on its age and the mean position within the condition class revealed by inspection. For example, a 60 year old sewer was found to be still in good condition, cc4. It can therefore be said to be halfway between transition functions cc5 →cc4 and cc4 →cc3. By finding the point at which a 60 years old sewer exists halfway between these transition functions the aging speed can be found, in this case approximately at 30% as shown in Figure 7-2.

Without further information, the ages for entering subsequent condition classes are read on the 30% aging speed line; entering cc3 at 80 years and cc2 at 130 years. Assuming that rehabilitation is triggered upon entering cc2, the residual service life is approximately 70 years. It however takes another 40 years for the sewer to enter cc1, the ultimate condition requiring rehabilitation. During such long periods further inspections are needed to ensure that the assumed constant aging speed remains accurate. This should be done before entering cc2.

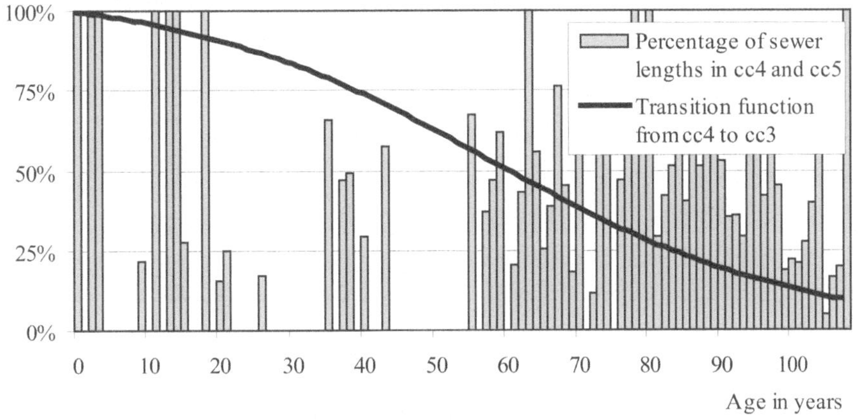

Figure 7-1. Calibrated Transition Function.
(Reprinted with permission from Baur and Herz, 2002)

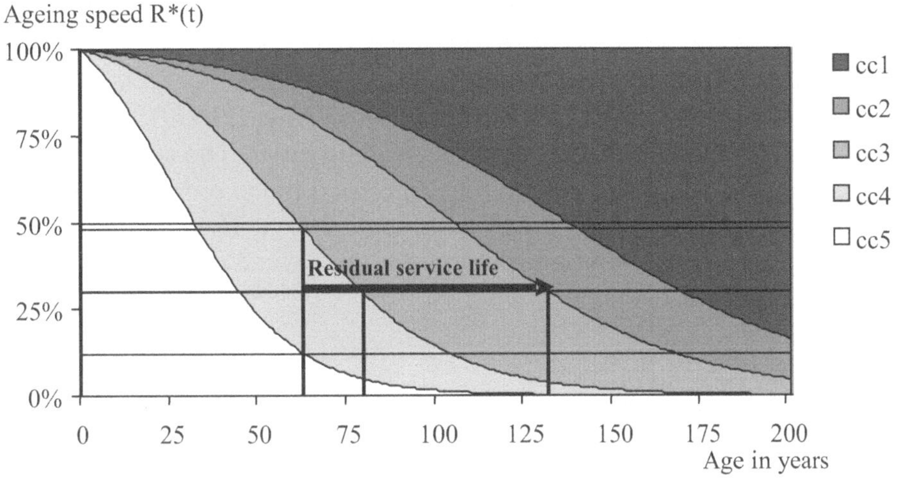

Figure 7-2. Aging Speed and Residual Service Life Estimate.
(Reprinted with permission from Baur and Herz, 2002)

The aging model allows the effect of various rehabilitation programs to be simulated and for potential future network conditions to be observed. If the results of various rehabilitation strategy simulations are compared to the consequences of the do-nothing option for unhindered network development, as shown in Figure 7-3, the benefits of different strategies can be calculated and compared. Consequently, rehabilitation programs and strategies can be developed in an interactive learning process, where alternatives are compared, evaluated and improved with respect to their costs and effects in both the short and long terms (Herz and Krug, 2000).

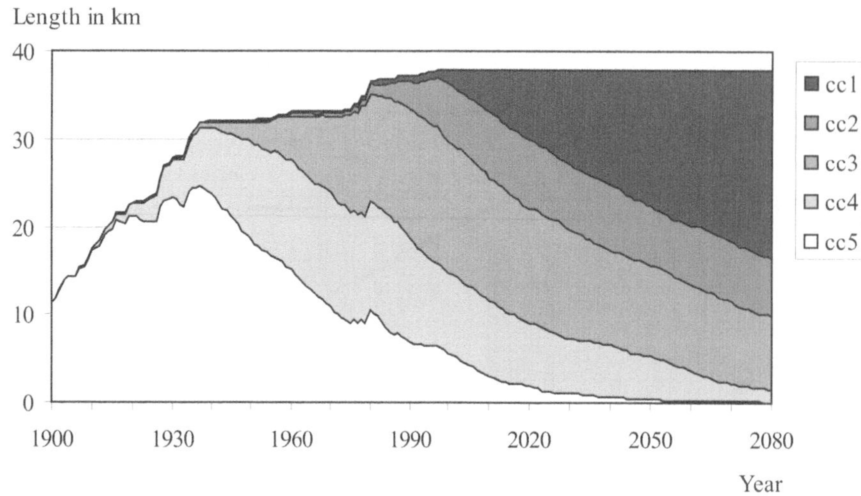

Figure 7-3. Condition Forecast for a Sewer Network.
(Reprinted with permission from Baur and Herz, 2002)

7.9 Statistical Methods – Ordinal Regression

The ordinal regression methods are similar to the cohort-survivor approaches and have become popular statistical models when dealing with a relationship between an integer-valued output and one or more explanatory variables (Johnson and Albert, 1999). The integer values are sometimes meaningfully ranked in increasing or decreasing order. The condition grading system of sewer is a typical example. Some common functions used in the ordinal regression models were logistic and probit functions (Tabachnick and Fidell, 2001). The maximum likelihood method is the commonly used calibration technique for the ordinal regression models (Johnson and Albert, 1999).

7.9.1 Summary of Approach

A summary of the ordinal regression approach to statistical methods for infrastructure deterioration is as follows

- In theory, this approach could be applied to any infrastructure assets that have recorded data describing the assets condition. However, it has been primarily applied to sewer pipes in the literature.
- Condition grade information is required (e.g., 1 = Good to 5 = Poor).
- This approach can only be applied to homogeneous groups (cohorts) of assets.
- Both asset-attribute and condition grade data must be available.
- Typical asset attribute data are: material, pipe length, installation date, diameter and soil type.
- No commercial tools are known, but models for sewer pipelines are well described in the literature.

7.9.2 Theoretical Background

Ordinal regression models assume that there is a latent continuous process such as deterioration of sewers and that the observed ordinal outcome arises from discretizing the underlying continuum into a number of ordered groups. A set of thresholds or cut-off values are used to separate these groups. The basic form of the model is shown below:

$$f(Y_j) = \theta_j + \beta_1 X_1 + \beta_2 X_2 + \ldots + \beta_k X_k \qquad (7\text{-}14)$$

where Y_j is the cumulative probability for the *j*th category or condition state for a pipe; β_1, β_2,..,β_k are the factor coefficients for *k* factors of a pipe; X_1, X_2 ,..,X_k are the predictors or pipe factors such as pipe size and pipe location; θ_j is the threshold for the *j*th condition state. The function $f(Y_j)$ is the function of the probabilities for condition grade and is the link between the random component on the left side of the equation and the systematic component on the right. Two commonly used link functions are:

- the Logit function: $\ln(\frac{Y}{1-Y})$ and
- The Probit function: $\Phi^{-1}(Y)$.

Φ in the Probit function is the standardized Normal distribution and can be obtained from tables in the literature (Thoft-Christensen and Baker, 1982). Probit and Logit models are reasonable choices when the changes in the cumulative probabilities for condition grade are gradual. If there are abrupt changes, other link functions should be used. The complementary log-log function $\ln[(-\ln(1-Y)]$ may be a good model when the cumulative probabilities increase from zero (0) fairly slowly and then rapidly approach 1. If the opposite is true, namely that the cumulative probability for lower scores is high and the approach to 1 is slow, the negative log-log function $-\ln[-\ln(Y)]$ may better describe the data. The ordinal model is calibrated by using the maximum likelihood method for the sample data of CCTV inspected pipes, as described in Crowder et al. (1991).

7.9.3 Practical Considerations

Ordinal regression models were developed in several studies to identify whether a particular sewer is in good or bad condition. Examples are described by Davies et al. (2001b), Ariaratnam et al. (2001) and Koo and Ariaratnam (2006) for studies of structural deterioration of combined sewers in UK, Canada, and U.S. respectively. Pohls (2001) documents an investigation of sewer blockages in Australia. However, these studies were focused on the investigation of factors that affect the structural deterioration and lacked goodness-of-fit tests comparing the results of the deterioration models with the observed data.

7.9.4 Examples of Application

Ariaratnam et al. (2001) developed a logit model for prioritizing sewer inspection for the City of Edmonton in Canada. A sample of CCTV-inspected pipes was used to calibrate the model and then predict the future condition of sewers. For ease of interpretation, their logit model reduced the 5 condition ratings that were reported into a binary condition state of 'good' (condition 1-3) and 'bad' (condition 4-5).

The model was calibrated using a sample of CCTV inspected data. The calibration method maximized the likelihood function to observe the sample, as in Equation 7-15.

$$Maximize = \prod_{n=1}^{N} Y_n^{\alpha} (1 - Y_n)^{1-\alpha} \quad (7\text{-}15)$$

where: α is 0 if observed condition is zero (good), otherwise is 1 (deficient)

Y_n is the probability that pipe n is in condition 1

Each sewer had pipe factors including diameter, age, waste type, material type and depth of cover. Among these factors, only three factors were found statistically significant and were used in the following formula to find the probability of a deficient sewer:

$$prob(deficient\ sewer) = \frac{\kappa}{1+\kappa} \quad (7\text{-}16)$$

$$\kappa = e^{-0.0839+0.0259\,AGE-0.0039\,DIAMETER+2.2191\,WASTE-TYPE} \quad (7\text{-}17)$$

Davies et al. (2001b) conducted a similar study in the UK, in which condition records were reduced to "poorest condition" (condition grade 5) and "not poorest condition" (condition grades 1-4), and the result was subject to a regression analysis. In contrast to Ariaratnam et al. (2001), a total of 19 pipe factors, as given in Table 7-3 below, were used in the logit model, with 10 found to be statistically significant (Table 7-4). The UK study indicated that the probability of imminent sewer collapse could not be predicted from the model and that a large number of significant variables affected the result. However, the collapse records used were found to be of unacceptable quality in terms of accuracy, completeness and consistency, which may explain this discrepancy.

Table 7-3. List of Input Factors in a Model of Sewer Pipe Collapse.
(Reprinted with permission from Davies et al., 2001)

Variable	Variable type	Levels (for categorical variables)	Data source
Sewer condition	Nominal categorical	0 = Grade 1/2/3/4 1 = Grade 5	CCTV
Sewer depth	Continuous quantitative		CCTV
Sewer use/purpose	Nominal categorical	0 = Combined 1 = Foul 2 = Surface water	CCTV
Sewer size	Continuous quantitative		CCTV
Sewer material	Nominal categorical	0 = Concrete 1 = Clay 2=Other	CCTV
Sewer pipe length	Ordinal categorical	0 = <0.9m 1 = 0.9-1.49m 2 = >1.49m	CCTV
Sewer location	Nominal categorical	0 = Main road – urban 1 = Main road – suburban/rural 2 = Light road 3 = Footpath or verge 4 = Other	CCTV
Root intrusion	Nominal categorical	0 = No root intrusion 1 = Root intrusion (fine, mass or tap	CCTV
Infiltration	Nominal categorical	0 = No infiltration 1 = Infiltration (dripper, seeper, runner or gusher)	CCTV
Debris	Nominal categorical	0 = Debris (grease or silt) 1 = Debris (grease or silt)	CCTV CCTV
Soil fracture potential	Ordinal categorical	0 = Very low 1 = Low 2 = Moderate 3 = High 4 = Very High 5 = High* (alluvial)	Digital soil maps
Soil corrosivity	Ordinal categorical	0 = Non aggressive 1 = Slightly aggressive 2 = Moderately aggressive 3 = Highly aggressive 4 = Very highly aggressive	Digital soil maps
Goods vehicle flow	Discrete quantitative		LRC traffic data
Vehicle flow	Discrete quantitative		LRC traffic data
Bus flow	Discrete quantitative		LRC traffic data
Road classification	Nominal quantitative	0 = 'A' road 1 = 'B' road 2 = Unclassified	LRC traffic data
Burst history	Continuous quantitative		Burst database
GW regime	Nominal categorical	0 = In vicinity of invert 1 = Generally below invert 2 = Indeterminate 3 = Tidal influence	TW borehole log
Property age	Ordinal categorical	0 = Before 1900 1 = 1900-1924 2 = 1925-1949 3 = 1950-1974 4 = 1975-1999	Property age data

Table 7-4. List of Statistically Significant Factors in Model of Sewer Pipe Collapse.
(Reprinted with permission from Davies et al., 2001b)

Variable/factor	Change in deviance	Degrees of freedom	*P*
Debris	381.68	1	<0.01
Sewer pipe length	257.88	2	<0.01
Sewer size	137.20	1	<0.01
Sewer use	85.36	2	<0.01
Soil fracture potential	83.17	4	<0.01
Soil corrosivity	29.76	4	<0.01
Sewer location code	39.38	4	<0.01
GW regime	12.87	3	<0.01
Sewer material	7.70	2	0.02
Bus flow	12.35	1	<0.01

In a similar study conducted in the U.S., Koo and Ariaratnam (2006) developed a deficiency logit model for prioritizing sewer inspection for the City of Phoenix. As with the Canadian and UK studies, they also used a sample of CCTV-inspected pipes to calibrate their model. In this case, only three pipe factors (age, maximum flow velocity and cumulative flow) were used. For model testing, sample data was divided into separate groups according to three sub-classes of input factors as shown in Table 7-5.

Table 7-5. Sub-class of Three Input Factors.
(Reprinted with permission from Koo and Ariaratnam, 2006)

Parameter	Group 1	Group 2	Group 3
Age	14 – 28 years	29 – 35 years	>35 years
Maximum Velocity	<2.5 ft/s	2.5 – 6.0 ft/s	>6.0 ft/s
Cumulative	< 11,000 gal/min	11,000 – 26,000 gal/min	>26,000 gal/min
Flow	< 183.33 gal/h	183.33 – 433.33 gal/h	> 433.33 gal/h

Cumulative flow was found to be statistically insignificant during model calibration. The predicted and observed numbers of deficient sewers in each group were compared against each other as shown in Figure 7-4 below.

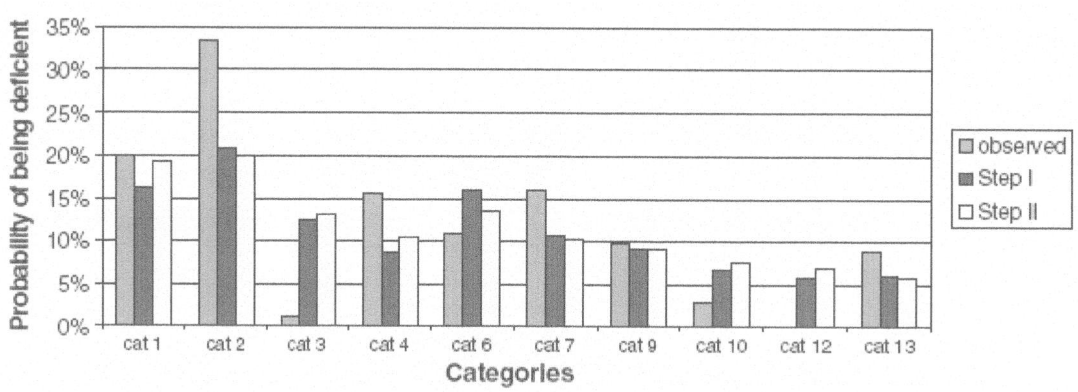

Figure 7-4. Observed/Predicted Frequency of Sewer Deficiency.
(Reprinted with permission from Koo and Ariaratnam, 2006)

In Figure 7-4, 'Step 1' corresponds to the use of all three factors in model calibration and 'Step 2' means only the two significant factors (age and maximum velocity) were used. They achieved 13.45% and 12.48% for the maximum differences between the observed and predicted number in Step 1 and Step 2. However, Koo and Ariaratnam (2006) did not indicate whether the testing data were excluded from model calibration. As such, there is no clear indication of the goodness of fit for the model in this case.

7.10 Statistical Models – Markov Chain Approaches

An alternative technique to modeling the change in condition for infrastructure assets is the application of Markov Chain models. As with cohort and ordinal regression approaches, these are inspection-based models and rely on asset condition indicators. However, rather than aging and deterioration through a continuous increase in recorded failure rate, Markov Chain models assume that deterioration occurs by a series of 'jumps' from one observed condition state to a more severe state over time. The probability of a jump occurring depends on asset age and operating environment. As with statistical models for failure rate forecasting, inspection-based models are based on assumed probability distributions that describe the underlying stochastic deterioration process. Model calibration is again conducted by matching predictions of condition state to observations from CCTV inspection.

7.10.1 Summary of Approach

A summary of the Markov Chain approach to statistical models for infrastructure deterioration is as follows:

- Commonly used for modeling deterioration of infrastructure facilities.
- In theory, this approach could be applied to any infrastructure assets that have recorded data describing the assets condition. However, it has been primarily applied to bridges, pavements, sewer and stormwater pipes.
- Simplified approaches deal with pipe cohorts or more advanced approaches deal with individual assets.
- Has the ability to model the condition change of a pipe population with snapshot data or individual asset (regular) data.
- Models for sewer pipelines are well described in the literature and underpin the Computer Aided Rehabilitation of Sewers (CARE-S) software tool.

7.10.2 Theoretical Background

According to Morcous et al. (2002), the Markov chain theory is still the most frequently used method in statistical models for the deterioration of infrastructure facilities. The basic idea of the Markov chain theory is that the prediction of the future condition over a unit time depends only on the current condition, regardless of its history. For example, a pipe which is in condition state 1 at the current time has a series of probabilities (namely P_{11}, P_{12}, and P_{13}) to remain unchanged or transition to one of the poorer condition states, 2 and 3 respectively, at the next period. Thus, a transition matrix, **P**, can be established for three possible condition states as shown in Equation 7-18:

$$\mathbf{P} = \begin{bmatrix} P_{11} & P_{12} & P_{13} \\ P_{21} & P_{22} & P_{23} \\ P_{31} & P_{32} & P_{33} \end{bmatrix} \quad (7\text{-}18)$$

where the sum of row elements is always 1 and the size of the matrix is equal to the grading range of facility. The structure of the Markov model is based on Markov chain theory (Ross, 1972). The Markov chain operates in such a way that, whenever the process is in condition state i at year t, there is a probability P_{ij} that the process will move to state j at year $t+1$. The time interval of the Markov chain was chosen as one-year and was indexed using non-negative integers. The important property is that the probability P_{ij} depends only on the present state (state i), which means the process is independent of historical states (Ross, 1972). It can be seen from the transition matrix **P** that pipes deteriorate when $i > j$, pipe condition improves (i.e. receiving maintenance or repair) when $i < j$, and pipe condition remains unchanged when $i = j$. The transition matrix **P** can be extended for 5 or a larger number of condition states in the same manner.

If the initial condition state at year zero (0) expressed by $\mathbf{C}^0 = (C_1^0, C_2^0, C_3^0)$ is known deterministically or probabilistically, the probability distribution $\mathbf{C}^t = (C_1^t, C_2^t, C_3^t)$ of being in one of three condition states at year t can be computed using the Equation below, derived from the Chapman-Kolmogorov formula (Ross, 1972):

$$\left(C_1^t, C_2^t, C_3^t\right) = \left(C_1^0, C_2^0, C_3^0\right) \times \begin{bmatrix} P_{11} & P_{12} & P_{13} \\ P_{21} & P_{22} & P_{23} \\ P_{31} & P_{32} & P_{33} \end{bmatrix}^t \quad (7\text{-}19)$$

where: C_i^t is the probability being in the condition state I at year t

C_i^0 is the probability being in the condition state I at year 0

$\sum_{i=1}^{3} C_i^t = 1$

$t = 0, 1, 2, 3\ldots$

If the transition matrix **P** and the condition at year zero are known, then the future condition of a pipe at any time t can be calculated using Equation (7-19). However, due to the lack of longitudinal or regular data, it is impossible to estimate the transition matrix for each pipe. On the other hand, if a sample of inspected pipes is assumed to represent a 'homogenous' population, the transition matrix for the population can be estimated from that sample of pipes. In this case, the effects of factors around individual pipes such as pipe size and pipe location are ignored. This means that all pipes now have the same transition matrix and this transition matrix represents the behavior of the pipe population with regard to changes in the population condition over time.

The changes in condition of pipe populations can be established as follows. The initial condition of a pipe becomes the initial proportion of pipes in each condition. Since it is often assumed that the pipe was initially in condition 1 with certainty, this implies that the entire pipe population (100% of pipes) was also initially in condition 1. The proportion of pipes in each of the three possible conditions at any time t can be computed using Equation (7-19), where C_i^t means the proportion of pipes in condition i at year t. Since the condition of pipes at year zero is often assumed to be good (i.e., condition 1), the remaining task is to estimate the transition matrix, which is also called calibrating the Markov model.

The calibration of a Markov model is the task of applying the selected calibrating technique to a calibration dataset in order to estimate the model parameters or the transition probability. The calibration dataset is often randomly selected from a sample dataset and accounts for 70-80% of the sample population. The remaining dataset is called the test dataset and can be used to test the Markov model. The dataset is a sample of pipes that were CCTV-inspected and graded using a condition grading scheme.

Whether the dataset contains regular or snapshot inspection data is an important factor in selecting a suitable calibration technique. Regular inspection means a facility is subjected to consecutive inspections at every time interval. For example, every bridge in the U.S. is legally required to be inspected every two years during their lifetime (Madanat et al., 1995) which results in a database with regular or longitudinal data. Snapshot inspection means a facility receives only one (or at least irregular) inspections. Sewers in the U.S. and stormwater pipes in Australia are currently often subject to a snapshot inspection program (Kleiner, 2001; Kathula, 2001; Wirahadikusumah et al., 2001; Baik et al., 2006); i.e., a small portion of pipes is selected for inspection according to age or some other selection process. The purpose is to take a snapshot view of network or asset condition at the time of inspection.

Having recognized that a snapshot dataset is often available for sewers, researchers have attempted different calibration techniques for the Markov model in order to: (1) predict the average condition changes of the pipe network over time, and (2) predict the condition change of individual pipes.

The Bayesian Markov Chain Monte Carlo simulation was used in a study by Micevski et al. (2002) for stormwater pipes in Australia to predict the average condition changes of a pipe network. In this calibration, pipe factors such as pipe size and location were ignored because all pipes were considered to come from a homogenous population. Furthermore, the transition probabilities were estimated from their Posterior distributions as shown in Equation (7-20):

$$Posterior\ Distribution = \sum_{t=1}^{T}\sum_{j=1}^{3} n_j^t \log C_j^t \qquad (7\text{-}20)$$

where: C_j^t is the probability being in the condition state j at year t

n_j^t is the number of pipes in condition state j at year t with T is the oldest age in the dataset

The ordered probit technique has been used in a Markov model of bridges (regular data) (Madanat et al., 1995). This calibration technique was recently applied to estimate the transition probabilities of individual sewers (snapshot data) by Baik et al. (2006). In this calibration, the deterioration of a sewer n was modeled as shown in Equation (7-21):

$$Z_n = \boldsymbol{\beta}\mathbf{X}_n + \varepsilon \qquad (7\text{-}21)$$

where: Z_n is the latent deterioration of sewer n

$\mathbf{X}_n$ is the vector of pipe factors such as pipe size and location for the sewer n

$\boldsymbol{\beta}$ is the vector of pipe factor coefficients

ε is random error to follow a standard normal distribution $\Phi()$

The transition probability P_{ij} of sewer n (index n is ignored for ease of expression) was determined by using a set of threshold values $\boldsymbol{\mu}$ which is compared against the latent deterioration Z_n as shown in Equation (7-22):

$$P_{ij} = \Phi(\mu_{i(j-1)} - \boldsymbol{\beta}_i\mathbf{X}) - \Phi(\mu_{ij} - \boldsymbol{\beta}_i\mathbf{X}) \qquad (7\text{-}22)$$

The factor coefficients and threshold values for each condition i=1 *to* 3 can be estimated by maximizing the logarithm of likelihood function (7-23):

$$LL = \sum_{n=1}^{N}\sum_{j=1}^{3}\gamma_{nj}\Phi(\mu_{i(j-1)} - \boldsymbol{\beta}_i\mathbf{X}) - \Phi(\mu_{ij} - \boldsymbol{\beta}_i\mathbf{X}) \qquad (7\text{-}23)$$

where: N is the sample size

γ_{nj} is equal to 1 if the observed condition change is j and otherwise is zero

7.10.3 Practical Considerations

From a theoretical point of view, the Markov model is attractive because it can handle the uncertainty of condition change of pipe assets with consideration of maintenance, repair and condition assessment using CCTV inspection. However, its real life applications are often simplified and require careful attention due to lack of data. For example, the effects of maintenance and repair are not recorded for all infrastructure facilities which results in a simplified version of the Markov model. In this simplified model used in many studies (Abaza et al., 2004; Baik et al., 2006; Micevski et al., 2002; Wirahadikusumah et al., 2001), the lower triangle of the transition matrix P (such as P_{21}, P_{31} and P_{32}) is considered zero which means asset condition can not improve. Furthermore, regular inspection is not carried out for stormwater and sewer pipes, which results in lack of regular data for calibrating Markov models for individual assets. This means that for snapshot type of data, the Markov model cannot be applied to predict condition change of individual pipes. Despite this, some studies attempted to calibrate the Markov model for individual pipes by using optimization (Wirahadikusumah et al., 2001) and ordered probit (Baik et al., 2006) techniques. However, the Markov models were not fully tested due to lack of regular data. According to Wirahadikusumah et al. (2001) and Madanat and Ibrahim (1995), at least three consecutive inspection data sets should be collected for testing the Markov model.

For the snapshot type of data, the Markov model can still be applied to predict the condition changes of a pipe population or an assumed 'homogenous' group of pipes. The outcome of this Markov model, as outlined in the section below, can be used to estimate the annual budget requirement as described in a study by Tran (2007) on deterioration of stormwater pipes in Australia. For this purpose, the Markov model can be calibrated using Bayesian Markov Chain Monte Carlo simulation described by Micevski et al. (2002).

Depending on the type of infrastructure facilities, some studies assume that a facility can only experience a unit increase (or unit change) if its condition changes. For example, Madanat et al. (1995) considered that bridges can deteriorate from condition state 1 to 2 only due to the nature of bridge deterioration. However, Micevski et al. (2002) concluded from the fitness of their Markov model that the structural condition of stormwater pipes in Australia can experience multiple jumps in a condition change. Baik et al. (2006) also assumed multiple jumps in their Markov model for sewers in the U.S.

In theory, the transition matrix P may change over time; this is then called a non-homogenous (or non-stationary) Markov model. Wirahadikusumah et al. (2001) considered that the deterioration of sewers would be different at different ages and thus constructed a non-homogeneous Markov model for sewers. Micevski et al. (2002) argued that the use of a non-stationary transition matrix P would increase the uncertainty in estimating the transition probabilities and may reduce the performance of the Markov model. A fitness test in their study showed that their Markov model using a stationary transition matrix P was acceptable.

Finally, Mishalani and Madanat (2002) further classified Markov models into state-based and time-based models. The former estimates the probability of condition changes of an infrastructure facility over a unit time. The latter predicts the probability distribution of time spent to have a unit change of the asset condition. One critical problem with calibration techniques for state-based Markov models is that they require truly uncensored regular data, i.e., continuous inspection (Mauch and Madanat, 2001). An example is shown here to explain the censored and uncensored data (see Nelson, 1982 for more details).

If a new facility is observed during an experiment of two years and the facility still functions at the end of the experiment, a censored data point is obtained. If the facility failed during the experiment, an uncensored data point is obtained. The truly uncensored data requires the recording of the timing when a facility changes its condition state throughout its service life. However, this requirement is impractical when considering cost, even for facilities like bridges. If, for example, a pipe were inspected and graded as condition 3 at the age of 30 years, and suppose that the unit time for inspection is five-yearly, and the pipe was inspected again at the age of 35 years and graded 5. This means that the pipe either changed from condition 3 to 5 via 4 or jumped from 3 to 5, and the timing for those changes was not known. From these conceptual arguments it can be seen that continuous inspection must be applied in order to find out when condition 3 has changed to condition 5. In order to address this problem, the time-based Markov model was proposed. Examples can be found in the works by DeStefano and Grivas (1998), Mauch and Madanat (2001), Kleiner (2001) and Mishalani and Madanat (2002).

Although the time-based Markov models were found to be adequate for modeling deterioration of bridge decks, the assumption of unit change of facility condition is not appropriate for stormwater pipes as found by Micevski et al. (2002) who concluded that the deterioration of stormwater pipes might undergo a multi-state transition (e.g., jumping from condition 1 to condition 3 within a time step).

7.10.4 Examples of Application

Typically, Markov chain-based models assess the entire gravity sewer system and do not consider the local environment of each pipe or deterioration mechanisms of the various pipeline materials. The model works with representative samples of the sewer system selected using common characteristics such as pipe diameter, material, surcharging, water table, ground type, and traffic loading.

Pipe deterioration functions are created by comparing the Internal Condition Grades (ICGs) of sewers at different ages. The deterioration functions model the transition of sewers between ICGs as they age. The deterioration functions are calibrated using the initial and current ICGs of pipes. An iterative process is undertaken, refining the estimated deterioration function until the ICG predicted by the functions matches the observed profile.

The output from the model is a set of deterioration probability functions, which describe the proportions of sewers expected to move from one condition grade to the next over time. The predicted deterioration rates can then be used to set long-term budgets for investment in sewer rehabilitation.

Markov Model for Pipe Population

Tran (2008) developed stationary Markov models for deterioration of stormwater pipes. This model was calibrated using a sample of CCTV-inspected pipes. The input data are shown in the Table 7-6.

Table 7-6. Input Factors Used in the Study.
(Reprinted with permission from Tran, 2008)

Input factors	Description	Example
Pipe Identification	A unique number	987
Structural Condition	Ordinal (1-3)	2
Hydraulic Condition	Ordinal (1-3)	3
Pipe Size	Scale (225 to 1950 mm)	600
Pipe Age	Scale (0 to 65 years)	45
Pipe Depth	Scale (0 to 4.83 m)	2
Pipe Slope	Scale (0 to 22.85%)	5
Tree-count	Scale (1 to 22) (number of trees around pipe)	2
Pipe Location	Nominal (1-4)	1
Soil Type	Nominal (0-1)	0
TMI	Nominal (0-1)	1

Only 'structural condition' and 'pipe age' were used to prepare input data for the Markov model for pipe population. An example of input data is shown in Table 7-7.

Table 7-7. Input Data to the Markov Model.
(Reprinted with permission from Tran, 2008)

Age	Structural Condition (counts)			Total
	1	2	3	
0	0	0	1	1
2	0	2	3	5
10	7	0	0	7
17	0	0	11	11

The Markov model was then calibrated and the stationary transition matrix **P** was estimated as shown in Table 7-8 below. The estimated transition probabilities were used to construct a diagram showing the structural condition change of the pipe population (Figure 7-5).

Table 7-8. Estimated Transition Probabilities.
(Reprinted with permission from Tran, 2008)

		Future Condition State		
		1	2	3
Current Condition State	1	0.9455	0.0202	0.0343
	2	0	0.9996	0.0004
	3	0	0	1

The model developed by Tran (2008) assumes that all pipes come from a homogenous population (i.e. the effects of contributing factors were ignored). As shown in Figure 7-5, the conditions 1 and 3 continuously decreased and increased respectively over the years. If maintenance and rehabilitation actions were not carried out until the age of 120 years, the model predicts that about 80% of the pipe population would be in the poor structural condition.

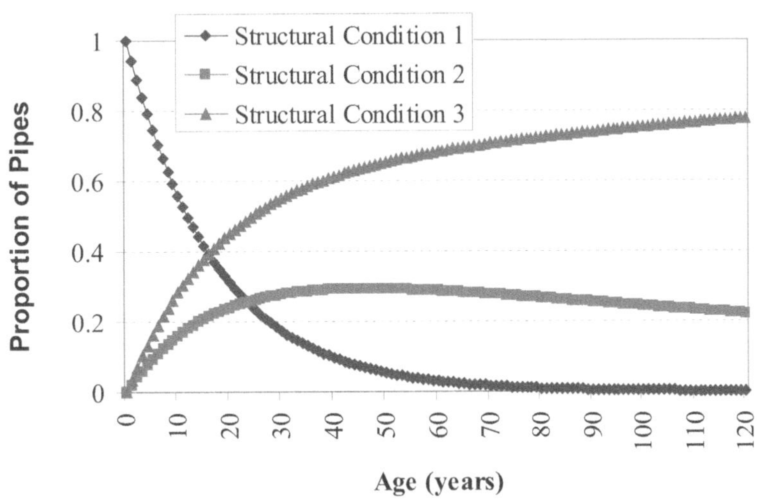

Figure 7-5. Condition Changes of Stormwater Pipe Population.
(Reprinted with permission from Tran, 2008)

Markov Model for Individual Pipes

As noted above, Baik et al. (2006) developed a stationary Markov model for prediction of structural condition of individual sewers in the U.S. The input data used in their study contained five factors as shown in Table 7-9; each sewer received only one inspection. The transition probabilities of each sewer were estimated (i.e. the model calibrated) by using the ordered probit technique, which was originally developed for calibration of the Markov model for bridges by Madanat et al. (1995). The ordered Probit technique relates the pipe factors (i.e. size, age) with the structural conditions (i.e. pipe deterioration) through a linear relationship and a set of thresholds.

The transition probabilities of a particular sewer given in Table 7-9 were estimated and then used to construct Figure 7-6. It can be seen from this Figure that, for example, if the condition of pipes at the age of 12 years is known as condition 1, then from the top left diagram in Figure 7-6, P_{11} has the largest value compared to that of P_{12}, P_{13} and so on, which means that the sewer is most likely to stay in condition 1 in the following stage (age of 13). However, since the sewer was inspected only once, it is not possible to test the consistency between the predicted values (condition 1) from the Markov model and the observed value (unknown condition). This is the major restriction of the Markov model for prediction of individual pipes with snapshot data.

Table 7-9. Input Factors used in the Study.

(Reprinted with permission from Baik et al., 2006)

Input factors	Description	Example
Pipe Identification	A unique number	987
Structural Condition	Ordinal (1-5)	2
Length	Between manholes in Feet	210
Size	Pipe diameters in Inches	8
Type of Material	Vitrified clay or PVC	45
Age	From installation to 2001	60
Slope	Between manholes	0.05

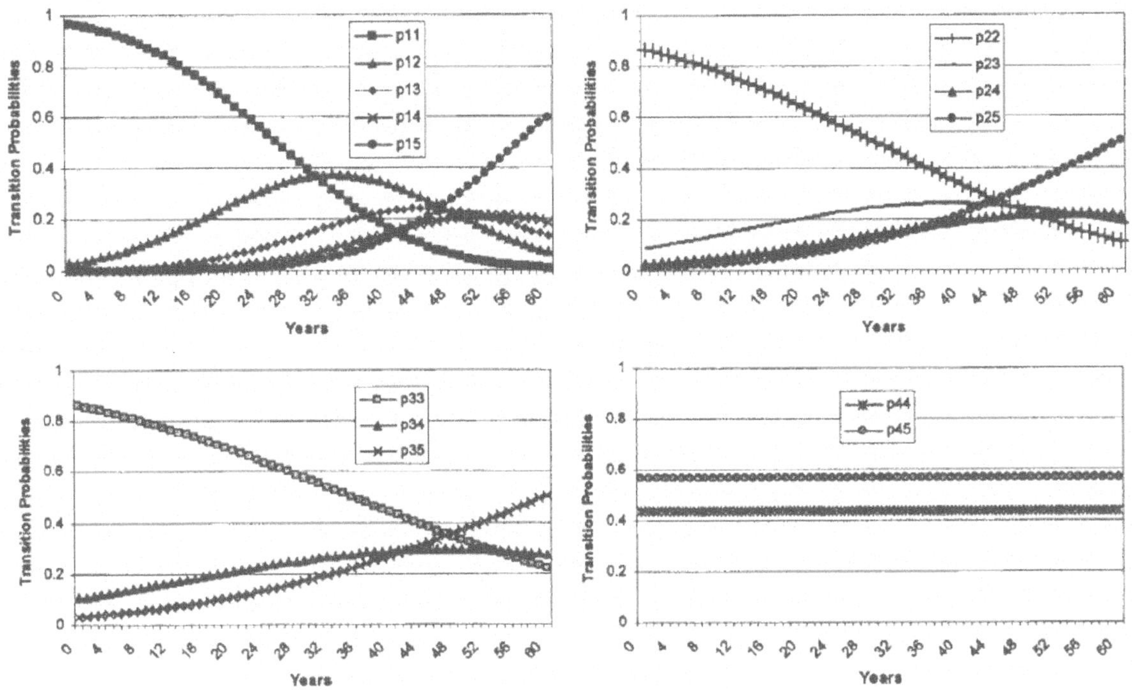

Figure 7-6. Transition Probabilities of an 8-Inch Sewer.
(Reprinted with permission from Baik et al., 2006)

Other examples of Markov models for individual pipes are documented by Le Gat et al. (1996) and Kleiner (2001).

7.11 Statistical Models – Bayesian Approaches

A summary of Bayesian approaches to statistical models for infrastructure deterioration is as follows:

- In theory, all infrastructure assets could be covered. However, models in the literature apply mainly to gas and water pipelines.
- With this approach, the probability of failure rates or condition states can be estimated for assets that belong to homogeneous groups (cohorts).
- Simple analysis requires asset length and failure data only. Detailed analysis requires relevant asset attribute information (i.e. material, size, age, etc.) for segregation into cohorts.
- No commercial tools are available, but a good description of model development and use is given by Kulkarni et al. (1986).

7.11.1 Theoretical Background

Bayes' Theorem was developed in the 18th century by the Reverend Thomas Bayes. It provides a relationship between the likelihood that an initial theory is correct after the addition of new data and the previous perceived likelihood, before the new data were added. Bayesian statistics enables "prior estimates" of the condition of the assets to be combined with any available site inspection data to make the best possible inference about the condition of the entire asset group.

Kulkarni et al. (1986) developed a maintenance system for cast iron gas pipelines using Bayesian statistics. The basic premise of the model was that the probability of failure in a pipe segment can be related to three factors:

1. The system-wide probability of failure P_f
2. The probability of observing specified characteristics on a segment that had failed $P_{c/f}$
3. The probability of observing the same characteristics on a segment that has not failed $P_{c/nf}$.

Following Bayes' Theorem, Kulkarni et al. (1986) proposed the following relationship:

$$\text{Prob}[\,failure \mid given\ specified\ characteristics] = \frac{P_{c/f}P_f}{P_{c/f}P_f + P_{c/nf}(1-P_f)} \qquad (7\text{-}24)$$

If the Prob[*failure given specified characteristics*] is close in value to P_f (= system wide failure probability), then this particular set of characteristics has little diagnostic capability. If the value is significantly larger or smaller than P_f, then the presence of this set of characteristics can "explain" a relatively high (or low) breakage rate in the gas pipelines. The system-wide average failure probability is calculated as:

$$P_f = \frac{\sum_{i=1}^{N} \text{Total number of segments with at least one break in year } i}{\sum_{i=1}^{N} \text{Total number of segments in year } i} \qquad (7\text{-}25)$$

where N is the total number of years accounted for in the database. The probability of observing specified characteristics in a pipe that has failed is:

$$P_{c_j|f} = \frac{\sum_{i=1}^{N} \text{Total number of segments with characteristics } c_j \text{ show at least one break in year } i}{\sum_{i=1}^{N} \text{Total number of segments that show at least one break in year } i} \qquad (7\text{-}26)$$

where c_j is the set of characteristics, also called the condition state. Condition states were defined by combinations of individual characteristics (Kulkarni et al., 1986). For example, a condition state of a pipe segment can be: age 50 years, 150mm diameter, 100 m long in clay soil, etc. After identifying the condition states c_j, for which P_f/c_j is significantly different from P_f, the expected number of failures in one year on a segment can be calculated as

$$\begin{aligned} &E[n \mid c_j] = \text{Prob}[\text{at least one failure in one year} \mid c_j] \times \\ &[\text{expected number of failures in one year on a segment} \\ &\text{with at least one failure in condition state } c_j] \end{aligned} \qquad (7\text{-}27)$$

The first term on the right hand side of Equation (7-27) is calculated simply from Equation (7-24). However, as described by Kleiner and Rajani (2000), the second term depends on the segment length. This dependence can be estimated by an appropriate regression analysis of number of failures versus segment length. Kulkarni et al. (1986) validated the above procedure on two sets of data and reported good agreement between the predicted and observed failure rates.

A similar Bayesian methodology has been applied in the UK by Water Research Centre, who developed a 'Dynamic Asset Condition Model' (DACM), a software application that could be used to forecast asset deterioration on the basis of an initial estimate of the deterioration rate of an asset group plus data from ongoing inspections. The model provides a condition estimate based on existing knowledge of a number of the assets in the group. By including estimates of the deterioration rate, the model can also predict the future condition of the assets. The use of Bayesian statistics allows the condition and deterioration rate estimates to be steadily improved as more inspections are carried out, providing a permanently up-to-date view of the investment requirement.

A Bayesian approach to modeling failure rates in water pipelines is also described by Watson et al. (2001). Although no precise details are supplied, the Bayesian model of Watson et al. (2001), combines engineering knowledge (in the form of expert opinion and beliefs about failure rates) with the failure rate data at hand to provide a formal estimate of the likely breakage rate distribution. According to Watson et al. (2001), this is written as:

$$P(\lambda|data) \propto P(\lambda).P(data|\lambda) \qquad (7\text{-}28)$$

where λ is the failure rate based on belief. However, no details are supplied on how Equation (7-28) is calibrated and used to forecast failure rates in practice (Watson et al, 2001).

7.11.2 Practical Considerations

Bayesian approaches to modeling infrastructure asset failures are based on simple and robust mathematical principles. As with other statistical techniques, assets are usually partitioned in homogeneous groups (cohorts) and the criteria for this segregation must be carefully defined. To calibrate and apply the models, the failure ratios in these groups are systematically compared and evaluated. As noted by Kleiner and Rajani (2000), the model of Kulkarni et al. (1986) is "flat" with respect to time and does not consider the time of breakage occurrence, just the empirical number of breaks to the time of analysis. Although no details are supplied, Watson et al. (2001) indicate that their Bayesian approach solves this limitation.

For a simple Bayesian analysis, only asset lengths and failure history are needed. Following Kulkarni et al. (1986), a detailed analysis requires partitioning of assets into an extensive list of condition states. As noted above, the appropriate criteria for this portioning must be available and chosen carefully.

7.11.3 Examples of Application

As described above, Kulkarni et al. (1986) provide a detailed description of a Bayesian diagnostic model for cast iron gas pipelines. The model was subsequently combined with failure cost data and developed into the Cast Iron Maintenance Optimization System (CIMOS). The Water Research Center in the UK produced a Dynamic Asset Condition Model, which was underpinned by the use of Bayesian statistics to update deterioration rates in assets based on subsequent inspections.

7.12 Physical Probabilistic Models – Monte Carlo Simulation

In contrast to purely statistical approaches, physical models are based on the actual degradation and failure processes that occur in service. Examples are the modeling of time-dependent pitting corrosion in metallic pipes, cement leaching and strength loss in cement-based assets and fracture from inherent defects in plastic pipes. While these physical models are developed and calibrated using small samples on laboratory scale (Davis et al., 2001; Rajani and Makar, 2000), they can be extrapolated to the network-wide scale through the incorporation of uncertainty in key variables. One technique is to represent key variables (i.e. degradation rate and operating loads) by appropriate probability distributions and use Monte Carlo (MC) simulation to generate projected service lifetimes for large numbers of assets. In this way, the details of physical degradation processes are retained but realistic uncertainty in this degradation is also accounted for. Outputs from these MC simulations are forecasts of time-dependent failure probability or estimated network wide failure rates, both of which can assist risk assessment and asset planning. Recent examples of physical probabilistic models are those developed for PVC and Polyethylene water pipes (AwwaRF, 2005a; Davis et al., 2007).

7.12.1 Summary

A summary of the Monte Carlo Simulation approach to physical probabilistic models for infrastructure deterioration is as follows:

- Primarily developed for pressure pipelines for water, oil and gas transportation.
- Models can estimate failure probability for individual assets or average failure rates at network-wide scale.
- For a particular material, models require asset attribute information, operating loads (i.e. internal pressure and external loads for buried pipelines) and condition data. Condition data depend on the particular material (i.e., corrosion damage in metals, residual strength in cement-based and defect size in plastics).
- All different asset types and loading conditions must be described for a model to be applied across the entire asset set.
- No commercial tools are available, but details of development, benchmarking and application of models are reported in the literature (Sadiq et al., 2004, AwwaRF, 2005a, Davis et al., 2007).

7.12.2 Theoretical Background

In the absence of historical failure or inspection data, physical probabilistic models can be developed to forecast failure probability. In contrast to traditional statistical approaches, such models are based on the actual degradation and failure processes that occur in an asset under in-service conditions. However, they also attempt to account for the inherent uncertainty that is often associated with degradation and asset failure.

At the core of a physical probabilistic model is the underlying degradation and failure model, which is often based on load-capacity relationships. For example, Rajani and Makar (2000) proposed a fracture criterion for cast iron pipelines subjected to combined internal pressure and in-plane bending. The underlying physical model used fracture mechanics theory to relate the critical loading conditions to the extent of pitting corrosion that a pipe has experienced in service. If the rate of pitting corrosion is known, the model can be used to predict when fracture failure will occur in service (Rajani and Makar, 2000; Deb et al., 2003; Moglia et al., 2008). Similar models have been developed that predict failure from the rate of cement leaching and strength loss in Asbestos Cement pipes (Davis et al., 2008a) and fracture from defects in PVC and Polyethylene pipes (Davis et al., 2007; Davis et al., 2008b).

Having defined an underlying physical model, the next stage of model development is to account for realistic uncertainty that accompanies degradation and failure process. Although there are many physical models in the literature that predict failure from condition assessment measurements, an additional problem is presented by the inherent uncertainty of the degradation process. For example, although available information may indicate that a uniform soil type exists along a length of metallic pipeline, local variations in soil electrochemistry will lead to uncertainty in external surface corrosion rates (Moglia et al., 2008). Similarly, variations in the water chemistry in contact with the pipe inner surface will lead to inherently uncertain leaching rates in cement-based pipes (Davis et al., 2008). Variations in processing conditions can also lead to defect sizes that are inherently uncertain in plastic pipes (Davis et al., 2007 and 2008b).

Realistic uncertainty can be accounted for by representing model variables with appropriate probability distributions rather than single-valued (deterministic) quantities. For example, Moglia et al. (2008) demonstrated that the inherent variation in measured maximum corrosion rate along a cast iron pipeline was well represented by the two-parameter Weibull probability distribution. Rather than a single corrosion rate that is assumed to apply along the entire length, the adoption of the Weibull probability distribution allowed a mean value and a variance to be defined. The two-parameter Weibull distribution has also been applied to measured rates of strength loss in Asbestos Cement pipes (Davis et al., 2007) and measured defect sizes in PVC pipes (Davis et al., 2007).

A straightforward way to estimate probability functions for asset lifetime is to use Monte-Carlo simulation in conjunction with the underlying physical model of failure prediction. In Monte-Carlo simulations, a computer program repeatedly generates random values for model variables (such as corrosion rate and defect size) from their probability distributions. Each time a random value is generated, it is used in the physical model to predict a failure time.

The end result of a Monte Carlo simulation is a set of predicted lifetimes, the number of which corresponds to the number of trials in the simulation. Monte Carlo simulations are run until some form of stopping rule is satisfied. For example, one commonly used rule is that the number of trials must be sufficiently large to ensure that the Standard Error of the Mean predicted lifetime is below a threshold value (Davis et al., 2008). The resulting set of predicted lifetimes is then fitted to an appropriate probability distribution, which is used to estimate failure probability over time. For modeling the structural reliability of pressure pipes, Monte-Carlo simulations are simple to set up and model detail and computer processing time are not usually limiting factors (Sadiq et al., 2004; Moglia et al., 2008). The steps required for Monte Carlo simulation of the physical failure model are outlined below:

1. Set up a hypothetical population of assets (or trials)
2. Randomly assign variables for the underlying physical failure model in each trial based on their probability distribution
3. Increment time
4. Calculate load experienced by each asset

5. Calculate the resistance for each asset
6. Check all asset segments for failure using the underlying physical model
7. Record all asset failures before returning to step 4
8. Estimate the expected lifetime and standard deviation from the predicted lifetimes in all trials
9. Calculate the standard error of the mean (SEM) lifetime and check that the number of simulated pipes assigned in step 1 is sufficient

In step 9, it must be demonstrated that the number of trials simulated (i.e. the number of hypothetical pipes chosen in step 1) is sufficient. One commonly used approach is to base this decision on the *standard error of the mean* (*SEM*) statistic:

$$SEM = \frac{\sigma}{\sqrt{N}} \tag{7-29}$$

where σ is the standard deviation of the variable of interest (in this case the predicted asset service lifetime), and N is the number of trials in the simulation. A proposed rule of thumb for Monte Carlo simulation is that the number of trials must be sufficiently high that the SEM statistic is less than 1% of the mean predicted lifetime (Schuyler, 1997). However, Davis et al. (2008) have shown good reproducibility in simulation of failure in Asbestos Cement pipelines when SEM is between 1% and 2% of mean predicted lifetimes

In contrast to First-Order Second Moment (FOSM) methods, Monte Carlo simulations allow the details of the physical failure mechanism to be preserved with no linearization required. Also, non-Normal distributions for stochastic variables can readily be accommodated in the limit state function. Figure 7-7 shows an example of output from repeated Monte Carlo simulations of a physical failure model for Asbestos Cement pipes.

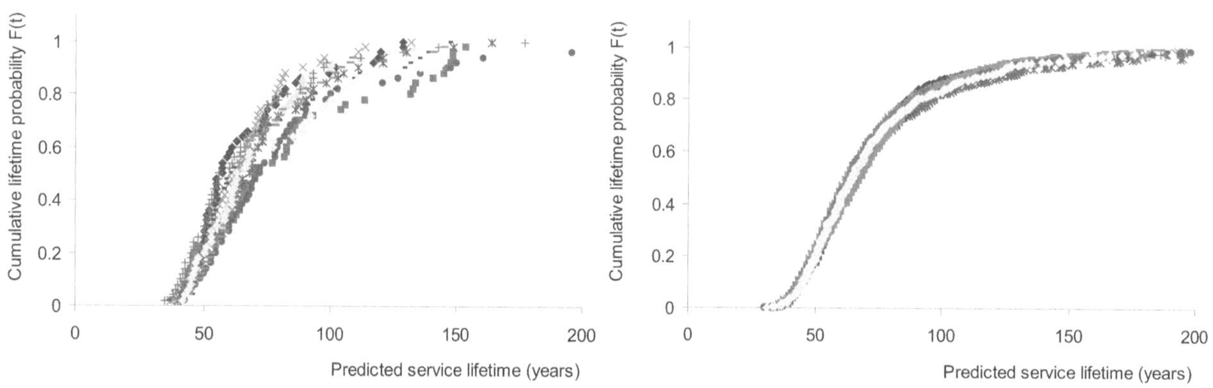

a) 100 Trials *b*) 500 trials

Figure 7-7. Empirical Cumulative Probability Distribution.
(Reprinted with permission from Davis et al., 2008)

Having set up a Monte Carlo simulation of service lifetime, empirical cumulative probability distributions for asset service lifetime can be generated as shown by the example in Figure 7-7. However, in order to use results from the simulation to estimate failure probability and failure rates, the simulated lifetime data must first be represented by an appropriate probability distribution. As outlined previously, a number of candidate distributions are available for modeling simulated lifetime data. Of these, the two-parameter Weibull distribution (Davis et al., 2007; Moglia et al., 2008) and the so-called Herz distribution (Davis et al., 2008) have been used to model lifetime data for water pipelines.

7.12.3 Practical Considerations

For those cases where flexibility in developing and changing the underlying physical model is imperative, Monte Carlo simulation is preferable. Within the scripting programming environment available today, Monte Carlo simulation is quick and easy to set up, and CPU time is no longer a limiting factor for many applications. Monte Carlo simulation bypasses the problem of computational complexity by using the brute force of modern computers. A major advantage of the Monte Carlo simulation methods is that it provides considerable flexibility about changing model assumptions.

Other than for metallic pipelines, there are no commercially available tools to measure degradation in-situ. As such, obtaining sufficient information to develop probability distributions for model variables can be costly and difficult to justify, especially for lower consequence pipelines. In the absence of measured data, distributions for model variables are assumed based on previous studies in the literature and, as such, failure rate forecasts can be inaccurate.

7.12.4 Examples of Application

While statistical-based models can be applied to older assets with sufficient historical failure or condition data, their applicability is limited for newer assets. Examples are those assets manufactured from relatively new materials (such as PVC or Polyethylene) and assets located in areas where relatively severe consequences would be incurred upon failure. In these cases, insufficient historical data exist for confident forecasts of future performance. An alternative approach is to develop physical probabilistic models that estimate the probability of asset failure over time. Examples of the application of physical probabilistic failure models are given in:

1. Rahman (1997): Monte-Carlo simulation to estimate the probability of leakage from nuclear piping and piping welds.
2. Sadiq et al. (2004): Monte Carlo simulation to estimate probability of brittle fracture in corroding cast iron pipelines.
3. Davis et al. (2007): Monte Carlo simulation to estimate probability of fracture in PVC pipes and network-wide failure rates in PVC pipelines.
4. Moglia et al. (2008): Monte Carlo simulation to estimate network wide failure rates in cast iron pipelines.
5. Davis et al. (2008): Monte Carlo simulation to estimate probability of longitudinal fracture in Asbestos Cement pipelines.

7.13 Physical Probabilistic Models – Structural Reliability-Based Methods

If suitable probability distributions can be selected for model variables, there are a number of approaches that accommodate this uncertainty in the underlying physical failure model. For example, methods such as the First-Order Second Moment (FOSM) and First-Order

Reliability Method (FORM) are elementary methods in Structural Reliability Theory. To apply these methods, the physical failure model is first re-formulated in the form:

$$M = L - R \tag{7-30}$$

where L represents the load experienced by the asset and R represents its ability to resist that load. M represents the limit state of the asset with failure predicted to occur when $M < 0$. Ahammed and Melchers (1997) demonstrate how L and R can be expressed in terms of an underlying physical model for combined corrosion and fracture of mild steel pipelines. Davis et al. (2007) demonstrated how a similar limit state equation can be used to predict crack initiation and fracture for PVC pipelines.

7.13.1. Summary

A summary of structural reliability-based methods applied to physical probabilistic models for infrastructure deterioration is as follows:

- Primarily developed for pressure pipelines for water, oil and gas transportation.
- Models can estimate failure probability for individual assets or average failure rates at network-wide scale.
- For a particular material, models require asset attribute information, operating loads (i.e. internal pressure and external loads for buried pipelines) and condition data. The type of condition data depends on the particular material (i.e., corrosion damage in metals, residual strength in cement-based materials, and defect size in plastics).
- All asset types and loading conditions must be described for a model to be applied across the entire asset set.
- No commercial tools are available, but details of development, benchmarking and application of models are widely available in the literature (Ahammed and Melchers, 1997; Rahman, 1997).

7.13.2 Theoretical Background

While underlying physical failure models can be readily developed, key variables that comprise the limit state equations are inherently uncertain. Structural reliability-based methods solve this problem by representing the (uncertain) limit state as a function of the inherently uncertain variables:

$$M = f(\overline{Z}) \tag{7-31}$$

M is now a stochastic limit state variable and $\overline{Z}$ is a random vector of the inherently uncertain basic variables that comprise the underlying physical model. The criterion for asset failure is now treated as a 'surface' in the variable 'space', with 'co-ordinates' corresponding to the probable values of the model variables. The probability of failure (i.e. Prob $M < 0$) is now calculated as the probability that the actual variables combine to produce a point in variable space that is beyond the failure surface described by Equation (7-31). To simplify the solution, FOSM methods assume that the stochastic limit state variable M is approximately Normally distributed and that the failure surface is a linear function of its variables. With these simplifying assumptions in place, the shortest distance from the failure region in variable space to the origin can be calculated and is known as the reliability index, β, (Thoft-Christensen and Baker, 1982).

Since the failure surface is linear and all model variables are assumed to follow the Normal distribution, the following relation exists between the probability of failure P_f and the reliability index β:

$$P_f = \Phi(-\beta) \tag{7-32}$$

where Φ is the standardized Normal distribution function and can be determined for different vales of β using tables in the literature (Thoft-Christensen and Baker, 1982). As degradation proceeds, the distance to the failure surface β decreases; this in turn increases the calculated probability of failure.

While FOSM relies on assumptions of Normal distribution and local linearity, approaches such as First-Order Reliability Method (FORM), Second-Order Reliability Method (SORM), and time-variant analyses are available that largely sidestep such limitations (Melchers, 1996).

7.13.3 Practical Considerations

While First-Order Second Moment (FOSM) reliability-based methods may offer analytical solutions to failure probability problems, they can also result in over simplification. For example, they require stochastic variables (such as defect sizes) to be Normally distributed and underlying failure models to be linear. In reality, these assumptions may not always hold. In fact, complex loading conditions experienced by infrastructure assets mean that linear failure is rarely experienced. As a consequence of all assumptions, realistic physical failure mechanisms can be lost in order to obtain a solution. However, extensions to reliability theory have produced First-Order Reliability Methods (FORM) and Second-Order Reliability Methods (SORM), which largely sidestep such limitations. However, for those cases where flexibility in developing and changing the underlying physical model is imperative, Monte-Carlo simulation is preferable.

Rahman (1997) used FORM/SORM as well as a Monte-Carlo simulation in a probabilistic fracture analysis of circumferential through-wall-cracked pipes subject to bending loads. In that study, Rahman achieved similar accuracy using FORM/SORM as with Monte-Carlo simulation, but the Monte-Carlo simulation used considerably larger amounts of CPU time.

7.13.4 Examples of Application

While statistical-based models can be applied to older assets with sufficient historical failure or condition data, their applicability is limited for newer assets. Examples are those assets manufactured from relatively new materials (such as PVC or Polyethylene) and assets located in areas where relatively severe consequences would be incurred upon failure. In these cases, there may be insufficient historical data exists for confident forecasts of future performance. An alternative is to develop physical probabilistic models that estimate the probability of asset failure over time. Examples of the application of physical probabilistic failure models are given in:

1. Ahammed and Melchers (1997): First-Order Second Moment (FOSM) method to predict probability of ductile rupture in corroding Mild Steel pipelines.
2. Rahman (1997): First-Order Reliability Method (FORM) to estimate probability of circumferential fracture in corroding mild steel pipelines.

7.14 Soft Computing Methods – Artificial Neural Networks (ANNs)

Artificial Neural Networks (ANNs) belong to the class of Artificial Intelligence modeling techniques. In contrast to statistical or physical probabilistic approaches, ANNs are data driven rather than model driven. ANNs predict output from input information in a manner that simulates the operation of the human central nervous system. Within the model, processing elements (or nodes) receive information that is then manipulated and passed on to other nodes in the network. ANNs consist of layers of nodes that provide a functional relationship between input information and predicted output. They are trained on historical data sets that demonstrate the actual relationship between input and output information and then tested on independent data.

7.14.1 Summary

A summary of the ANN approach to infrastructure deterioration is as follows:

- This approach has been used to model deterioration and failure in stormwater pipes (Tran et al., 2007) and water pipes (Achim et al., 2007).
- These models can be developed for individual asset or network-wide performance measures.
- Input data usually consist of all variables thought to influence service life. However, the precise number of input variables to an ANN depends on the particular problem to be solved. For pipeline applications, variables used include, size, age, operating loads and surrounding soil types.
- Typical number of training sets used to train ANNs for pipeline modeling is between 400 and 1000 (Tran et al., 2007; Achim et al., 2007).
- No specific tools are available for pipeline modeling. However, a number of mathematical modeling software packages are available that can be used with pre-requisite knowledge of Neural Network theory. These include SPSS, NeuralWare and MATLAB applications.

7.14.2 Theoretical Background

The human central nervous system is comprised of building blocks called neural cells or neurons. A neuron receives signals or inputs and produces a response or an output. In biological neurons, inputs and responses are electrical pulses. The input pulses are passed to the neuron through multiple channels (dendrites) and each dendrite has a contact point (synapse) that acts as a gate, which either permits or restricts inputs to a neuron. In this sense, a biological neuron acts as a function, which receives a set of inputs or parameters and produces an output.

ANNs provide a means of computation that is inspired by the structure and operation of the human central nervous system. ANNs consist of a number of processing elements, or nodes, that are interconnected. Within a particular ANN, these nodes are analogous to (and attempt to simulate the functions of) the neurons of the human central nervous system.

One of the most common ANN architectures is the so-called 'feedforward' network, in which nodes are arranged in layers: an input layer, one or more hidden layers, and an output layer. The input (x_i) from each node in a previous layer is multiplied by a connection weight, (w_{ji}). These connection weights are adjustable and are analogous to the coefficients of a statistical model. These weighted inputs are summed at the forward receiving node. This combined input (I_j) is then passed through a transfer function to produce the nodal output (y_i). Each nodal output is then passed forward to the next layer of nodes in the network and the process is repeated. The process is summarized in Equations (7-33) and (7-34) and illustrated schematically in Figure 7-8 below.

$$I_i = \sum w_{ji} x_i \qquad \text{Summation} \qquad (7\text{-}33)$$

$$y_i = f(I_j) \qquad \text{Transfer} \qquad (7\text{-}34)$$

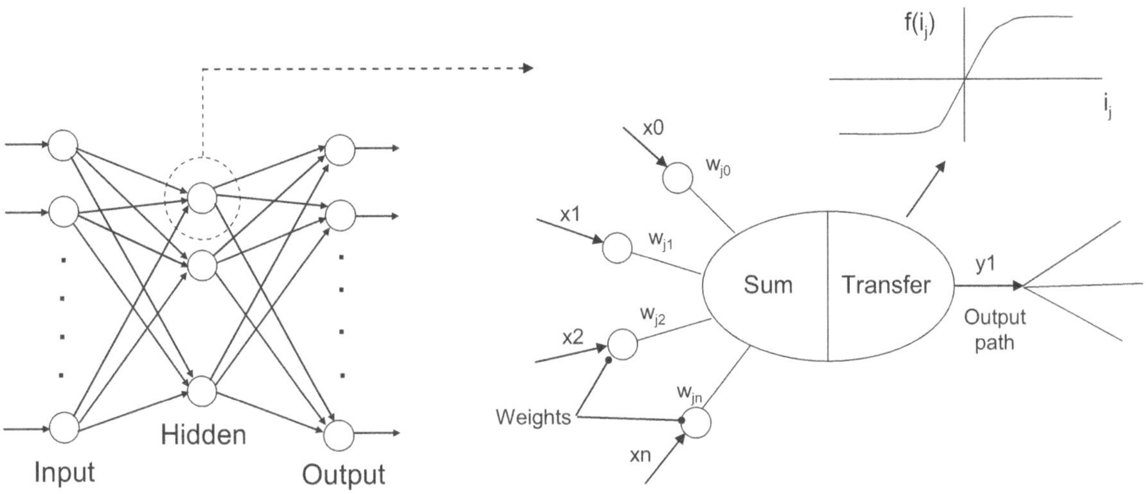

Figure 7-8. Structure and Operation of a Typical Feedforward Artificial Neural Network.
(Reprinted with permission from Maier and Dandy, 2001)

The purpose of an ANN is to capture the relationship between a historical set of model inputs and the corresponding outputs. This is achieved by repeatedly feeding in examples of input and output data and adjusting the model coefficients (i.e., the connection weights) in an attempt to minimize the error between the historical outputs and those predicted by the ANN. This process is referred to as 'training' the ANN, and a range of different optimization algorithms are documented in the literature to choose connection weights that minimize this error function. The most frequently used is the 'back-propagation' training process, in which the connection weights between nodes in the ANN are chosen based on the gradient descent method of optimization. The process is repeated in successive sets of training input/output data until certain stopping criteria are satisfied. For example, training may be stopped when a fixed number of training samples have been presented, when the training error has reached a predetermined value or when there is insignificant improvement in forecasts made on an independent test data set. A full description of the back propagation training process for feedforward ANNs is given by Maier and Dandy (2001).

The two most widely used transfer functions in feedforward ANNs are the sigmoid function and tangent function (Lou et al., 2001; Kingston et al., 2006), which are shown in Figure 7-9.

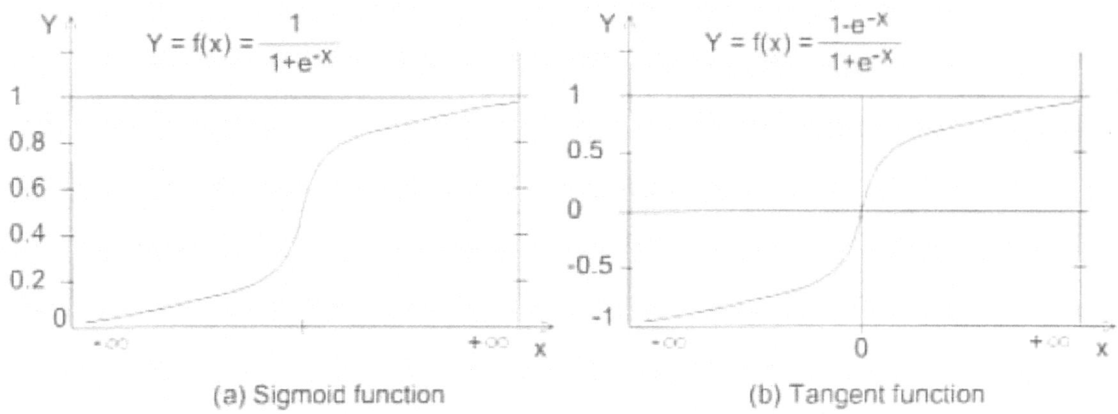

Figure 7-9. Transfer Functions Commonly used in Feedforward ANNs.
(Reprinted with permission from Tran, 2007)

These functions allow flexible non-linear mapping between input and output signals and, according to Bishop (1995), the use of either of these functions with a single hidden ANN layer can approximate any non-linear relationships in real world problems.

Having successfully trained an ANN using a set of historical input and output data, the model must also be validated. The generalization (or predictive) ability of an ANN is usually measured by the root mean squared error (RMSE) between the predicted and historical output values of an independent test set that was not part of the training data. By varying the number of training samples and calculating the RMSE on an independent data set, the generalization ability can be plotted during ANN training. Generally, a minimum RMSE is observed, beyond which further increases in training samples result in no further reduction in RMSE. This demonstrates successful training and signifies the highest level of generalization ability that can be attained for the particular problem.

7.14.3 Practical Considerations

While ANNs are being increasingly used to solve complex problems, they are also often treated as "Black Box" solutions. Data pre-processing, methods for determining adequate model inputs and the internal workings of ANNs are seldom considered during their application. In addition to expertise concerning the problem at hand, background knowledge in ANN development is also a pre-requisite.

According to Maier and Dandy (2001), during pre-processing of input data, any trends or heteroscedasticity (differing variances between model variables) should be removed prior to the training of an ANN. Another crucial issue is the number of model parameters or nodes included in the input layer. An insufficient number of inputs may result in difficulties reaching a minimum RMSE during training. Conversely, if too many model parameters are used in relation to the number of training samples available, the ANN may lose its ability to predict outcomes from independent input data. Another issue that should be considered is the number of training samples and the increment in nodal weights that are used during ANN training. Maier and Dandy (2001) demonstrate how an ANN can show divergent behavior and never reach an optimal training level if increments in nodal weights during the training process are too large.

7.14.4 Examples of Application

Tran et al. (2007) developed an ANN to predict deterioration and structural condition in stormwater pipelines. Input nodes to the network consisted of:

- Pipe diameter
- Age
- Burial depth
- Slope
- Location
- Number of surrounding trees
- Hydraulic condition (a qualitative indicator of the reduction in capacity)
- Soil Moisture Index (a surrogate indicator for soil movement)

The network consisted of a single hidden layer and model outputs were three different structural condition grades (1 – Good, 2 – Fair, 3 – Poor). For training data, these grades were based on expert assessment of CCTV inspection footage of the inside surface of stormwater pipes. After data cleaning and pre-processing, 583 data sets were available, which were randomly divided into training (437 data sets) and validation (146 data sets). Tran et al. (2007) demonstrate a performance rate (ratio of correct prediction to the number of input data sets) of 67% for the trained Neural Network model.

In another example, Achim et al. (2007) developed an ANN to predict the average failure rate (number of failures/length/year) for cast iron pipes in a water distribution network. Input nodes to the ANN were:

- Pipe diameter
- Year of construction
- Age
- Length
- Geographical x-coordinate
- Geographical y-coordinate

The ANN consisted of 2 hidden layers and a single output node containing the average failure rate. The ANN was trained using 1,024 cases where historical failure data was obtained between 1997 and 1999 and tested on 256 independent cases from this period. After training and validation, the model was compared to observed failure data recorded between 2000 and 2002. Regression coefficients indicated that the ANN gave better predictions than a conventional statistical model (Achim et al., 2007).

7.15 Soft Computing Methods – Fuzzy Logic

As with ANNs, the use of fuzzy logic theory to predict infrastructure deterioration is also part of the Soft Computing and Artificial Intelligence modeling category. Rajani et al. (2006) propose that the physicochemical processes that govern infrastructure deterioration are often not sufficiently well-understood to merit the development of a physicochemical model (e.g., based on mechanics, electrochemistry, and microbiology). Furthermore, converting deterioration observations to an infrastructure condition indicator is inherently imprecise and often involves subjective judgment (Rajani et al., 2006). Fuzzy logic-based techniques are able to incorporate engineering judgment and experience in a model development process. Specifically, they are proposed to be well suited to infrastructure problems in which:

- Data are scarce
- Cause-and-effect knowledge is imprecise
- Observations and model criteria are expressed in qualitative (narrative) terms such as 'poor', 'average' and 'good' condition)

As discussed by Kleiner et al. (2004), fuzzy logic-based techniques allow the propagation of these attributes through a model, therefore yielding more realistic results.

7.15.1 Summary

A summary of the fuzzy logic approach to modeling infrastructure deterioration is as follows:

- Previously used by Kleiner et al. (2006) to model deterioration and failure possibility in pre-stressed concrete cylinder pipelines and cast/ductile iron pipes.
- Input data usually consist of all deterioration variables (referred to as 'distress indicators') thought to influence condition. Some measure of their relative importance to overall condition is also required, which is obtained from expert opinion. The actual number of distress indicators used depends on the particular case being modeled.
- Data quantity depends on the particular case, but at least one observation set consisting of all relevant distress indicators is required for model calibration.
- No commercial tools for pipeline failure prediction are currently available, but a full description of model development and example software were developed as part of a project funded by AwwaRF (AwwaRF, 2005b). Mathematical analysis software packages such as *MATLAB* support the development of fuzzy logic-based models from first principles.

7.15.2 Theoretical Background

A fuzzy set describes the relationship between an uncertain quantity x and a membership function μ, which ranges between 0 and 1. Fuzzy sets are an extension of traditional set theory (in which x is either a member of set A or not) in that x can be a member of set A with a certain degree of membership, μ. Fuzzy techniques help to address deficiencies associated with vague (linguistic) information and are useful in propagating uncertainties from these sources through models. Although any 'shape' of membership function for fuzzy numbers is possible, the selected shape should be justified by available information. Generally, triangular fuzzy numbers (TFN) are used to represent linguistic variables and these are used in documented models for pipeline deterioration (Kleiner et al., 2006; Rajani et al., 2006).

To illustrate the concept of TFN, suppose that the age of a particular pipeline is defined by five fuzzy subsets (or numbers), where each subset represents an age grade: A_1 = *New*, A_2 = *Young*, A_3 = *Medium*, A_4 = *Old* and A_5 = *Very Old*. The schematic diagram in Figure 7-10 illustrates how these fuzzy subsets are defined using TFN.

Set	Qualitative scale	Triangular fuzzy number (TFN) representation
A_1	New	$(q_{1,1},q_{1,2},q_{1,3}) = (0,0,20)$
A_2	Young	$(q_{2,1},q_{2,2},q_{2,3}) = (0,20,30)$
A_3	Medium	$(q_{3,1},q_{3,2},q_{3,3}) = (20,40,60)$
A_4	Old	$(q_{4,1},q_{4,2},q_{4,3}) = (40,70,100)$
A_5	Very old	$(q_{5,1},q_{5,2},q_{5,3}) = (70,100,100)$

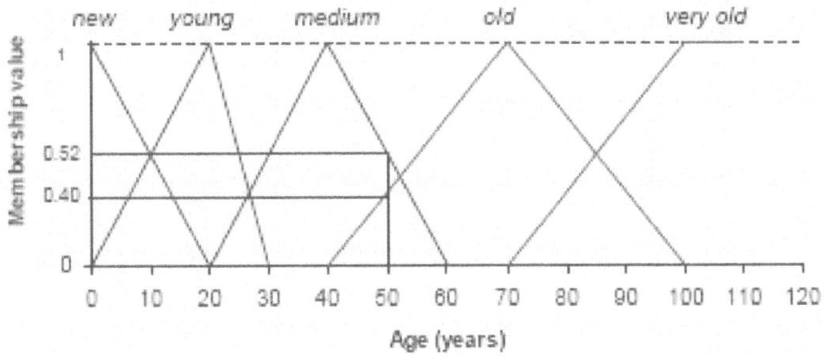

Figure 7-10. Use of Triangular Fuzzy Numbers (TFN) to Represent Age.
(Reprinted with permission from Kleiner et al., 2006)

In this example, a 50 year old pipeline has membership values $\mu_{A1} = \mu_{A2} = \mu_{A5} = 0$ and $\mu_{A3} = 0.52$ and $\mu_{A4} = 0.40$. The fuzzy set representing pipe age is therefore represented as the vector $A = (\mu_{A1}, \mu_{A2}, \mu_{A3}, \mu_{A4}, \mu_{A5}) = (0, 0, 0.52, 0.40, 0)$. Each element of the vector represents the membership of the age value to the linguistic age grade (from *new* to *very old*).

Extending the concepts of fuzzy numbers defined above, Rajani et al. (2006) have demonstrated how the condition of pipeline assets can also be represented as a fuzzy set. Building on this work, Kleiner et al. (2006) have demonstrated how fuzzy condition sets can then be combined with pipe age in a fuzzy logic-based approach to modeling deterioration rates. Both of these approaches are discussed in more detail below

Fuzzy Logic-based Approaches to Representing Pipe Condition

While in some cases, pipeline condition can be expressed in terms of clear numeric values (or grades), in many instances, translating observations to a measure of pipe condition involves subjective engineering opinion and vague (linguistic) information. It follows that this inherent uncertainty can be handled effectively using fuzzy logic-based approaches.

Rajani et al. (2006) describe a so-called "fuzzy synthetic evaluation" procedure in which observations of deterioration (termed 'distress indicators') in pipes are translated into a fuzzy set describing membership values to different condition ratings. The factors that contribute to pipe deterioration are organized in a two-level hierarchical structure. Level 1 consists of categories for distress indicators and level 2 consists of actual distress indicators in each category. A category reflects a specific pipe component (for example the external coating on a ductile iron pipe) and a distress indicator reflects the level of deterioration in that category and is a measurable quantity (for example the number of breaches to the external coating). Each distress indicator provides partial evidence to the condition of the specific component (i.e. external coating), while the category provides partial evidence to support the condition of the pipe. To illustrate these concepts, Rajani et al. (2006) document categories and distress indicators for cast/ductile iron pipes (Table 7-10).

Table 7-10. Distress Indicators for Cast/Ductile Iron Pipes.
(Reprinted with permission from Rajani et al., 2006)

Category (Level 1)	(j)	Distress indicator (Level 2)	(ij)	Comment
External coating (poly warp/tar/zinc)	1	Crack/tear/holiday	1,1	State of external coating will dictate how external corrosion is likely to encourage damage to the pipe.
External pipe Barrel / bell	2	Remaining wall thickness	1,2	Remaining pipe wall thickness is usually obtained from NDT tests or from spot exhumations and sand blasting samples.
		Graphitization (pit) aerial extent	2,2	Aerial extent as percentage of pipe diameter x unit length indicates the size of affected area. Severe graphitization may not always mean the pipe should have failed. In practice, graphitized area can still provide some resistance.
		Crack (pit) type	3,2	Circumferential cracks indicate some type of longitudinal movement has taken place. Longitudinal cracks occur due to low hoop resistance.
		Crack (pit) width	4,2	Crack width is another indicator of corrosion. Wide crack together with a deep pit will be more detrimental to the pipe than a narrow but shallow crack.
Inner lining / surface	3	Cement lining (epoxy) spalling (blistering)	1,3	Inner lining deterioration is often due to incompatible water chemistry or abrasion due to the presence of high water velocities and sediments.
		Remaining wall thickness	2,3	Occasionally CCTV scans can give estimates of internal corrosion pits when NDT tests are not done to get an overall picture of the pipe wall status.
		Tuberculation	3,3	Heavy tuberculation (blockage) can significantly reduce water delivery and produce red water condition.
Joint	4	Change in alignment	1,4	Changes in joint alignment indicate pipe susceptible to ground movement. Large changes can lead to leakage and eventually joint failure.
		Joint displacement	2,4	Joints can displace without undergoing joint misalignment and hence also an indicator of other forces at play.

For the hierarchical structure of categories and distress indicators defined by Rajani et al. (2006), M is the total number of categories, H_j denotes a particular category ($j = 1,\ldots,M$) and $H_{i,j}$ denotes a distress indicator i in category j. For example, $H_{1,2}$ represents distress indicator 1 in category 2. Each distress indicator is defined over a universal scale (in fuzzy logic terminology the 'universe of discourse'). The example in Table 7-11 shows the universe of discourse for distress indicators in cast/ductile pipes (Rajani et al., 2006).

Table 7-11. Universe of Discourse for Iron Pipes.
(Reprinted with permission from Rajani et al., 2006)

Category (level 1)	Distress indicator (level 2)	Universe of discourse
External coating (poly warp/tar/zinc)	Crack/tear/holiday	(no damage, slightly damaged, moderately damaged, highly damaged)
External pipe Barrel/bell	Pit depth	(pit depth <10%, 10%< pit depth <40%, 40% < pit depth <60%, 60% < pit depth < 80%, pit depth > 80%)
	Graphitization (pit) aerial exterior	(affected area < 20%, 20% < affected area < 40%, 40% < affected area < 60%, 60% < affected area < 80%, affected area > 80%)
	Crack (pit) type	(no cracks, circumferential, longitudinal – crown/invert, mixed (long + circum), longitudinal – spring line)
	Crack (pit) width	(no cracks, hairline, "dime" thick, >= "finger" thick)
Inner lining/surface	Cement lining (epoxy) spalling (blistering)	(no, yes)
	Remaining wall thickness	(pit depth <10%, 10%< pit depth <40%, 40% < pit depth <60%, 60% < pit depth < 80%, pit depth > 80%)
	Tuberculation	(diameter reduced by < 20%, 20% < reduced diameter < 40%, 40% < reduced diameter < 60%, 60% < reduced diameter < 80%)
Joint	Change in alignment	(no change, angle < 22.5° or Δ alignment < 5%, 6% Δ alignment < 10%, angle > 22.5 ° or Δ alignment > 11%)
	Joint displacement	(no displacement, displacement < 5 mm, 5 mm < displacement < 10 mm, displacement > 10 mm)

In Table 7-11, $H_{2,2}$ refers to the distress indicator 'Graphitization (pit) aerial extent' in the category 'External pipe barrel/bell'. The universe of discourse for this particular distress indicator is defined as *area < 20%; 20% < area < 40%; 40% < area < 60%; 60% < area < 80%; area > 80%*. Having defined the hierarchical structure for observed distress indicators, Rajani et al. (2006) propose that the translation to fuzzy condition ratings involves three steps: (1) fuzzification of raw data; (2) aggregation of distress indicators towards their respective categories and (3) aggregation of categories towards condition rating. These steps are presented in more detail below:

Step 1. Fuzzification of Raw Data

Rajani et al. (2006) assume that each individual distress indicator is associated with an underlying fuzzy set $C_{i,j}$, which is defined by the following 7 linguistic constants: *excellent, good, adequate, fair, poor, bad* and *failed*. Raw data for a given distress indicator is then fuzzified by mapping onto the fuzzy set $C_{i,j}$. For a given distress indicator, the membership functions to each element in $C_{i,j}$ are defined through expert opinion. For example, the fuzzy set that represents the distress indicator of graphitization over the extent *20% < area < 40%* ($H_{2,2}$ in Table 7-11) may be given by: $C_{2,2} = (0, 0, 0.7, 0.3, 0, 0, 0)$. This would indicate membership of 0.7 to the linguistic variable *adequate* and 0.3 to *fair*. Provided expert opinion is available, then the fuzzy set $C_{i,j}$ can be defined for all distress indicators in this manner.

Step 2. Aggregation of Distress Indicators to Categories

As described by Rajani et al. (2006), in Step 2 of the process, relative degrees of importance must be assigned to individual distress indicators in order to aggregate them towards their parent category. These degrees of importance are expressed as a weighting, which is assigned based on expert opinion. Let Wi,j denote a fuzzy set describing the relative weight of distress indicator i towards category j. Responses from expert opinion could be described in terms of the linguistic terms *extremely important, very important, important, moderately important* and *unimportant*. However, for further manipulation, Wi,j should be normalized to a sum of 1. Rajani et al. (2006) formally express the aggregation procedure as:

$$H_j = \lfloor W_{1,j} ... W_{i,j} ... W_{N_j,j} \rfloor \bullet \lfloor C_{1,j} ... C_{i,j} ... C_{N_j,j} \rfloor^T \tag{7-35}$$

In Equation (7-35), N_i is the number of distress indicators in category j. Each individual element $C_{i,j}$ is a 7-element fuzzy set that results from the fuzzification of raw data in Step 1. Therefore, $C_{i,j}$ is a $7 \times N_j$ matrix where N_j = number of distress indicators in category j. In contrast, each element $W_{i,j}$ is a scalar weighting (a single number). The subscript 'T' represents the transpose of the matrix $C_{i,j}$ and the '•' operator represents vector-matrix multiplication. The resulting vector H_j is therefore a 7-element fuzzy set representing the aggregated contribution of category j towards the final condition rating, which is expressed in terms of linguistic constants excellent, good…etc.

The application of Equation (7-35) is best illustrated through a hypothetical example. Referring back to 7.10, it is assumed that the 7-element fuzzy set H_2 for category 2 – 'External pipe barrel/bell' is required. The aggregation is achieved as outlined below. From Table 7-11, the universe of discourse for category 2 is defined as: *pit < 10%; 10% < pit < 40%, 40% < pit < 60%, 60% < pit < 80%; pit > 80%*. Assume that, after a process of expert survey, the following importance level was assigned to the first distress indicator pit < 10%: $C_{1,2}$(pit< 10) = [1, 0, 0, 0, 0, 0, 0]. As outlined by Rajani et al. (2006), the elements of $C_{i,j}$ for each distress indicator are membership values to the linguistic constants *excellent*, *good, adequate, fair, poor, bad* and *failed.* Assume that expert opinion allows $C_{1,2}$ to be defined for all distress indicators in the category:

$C_{1,2}$(pit < 10) = [1, 0, 0, 0, 0, 0, 0]

$C_{2,2}$(10 < pit < 40) = [0, 0.3, 0.7, 0, 0, 0, 0]

$C_{3,2}$(40 < pit < 60) = [0, 0, 0.3, 0.7, 0, 0, 0]

$C_{4,2}$(60 < pit < 80) = [0, 0, 0, 0.3, 0.7, 0, 0]

$C_{5,2}$(pit > 80) = [0, 0, 0, 0, 0.3, 0.7, 0]

Following Equation (7.35), this allows the matrix $[C_{1,2}, C_{2,2}, C_{3,2}, C_{4,2}, C_{5,2}]^T$ to be defined as:

$$\left[C_{i,2}\right]^T = \begin{bmatrix} 1 & 0 & 0 & 0 & 0 \\ 0 & 0.3 & 0 & 0 & 0 \\ 0 & 0.7 & 0.3 & 0 & 0 \\ 0 & 0 & 0.7 & 0.3 & 0 \\ 0 & 0 & 0 & 0.7 & 0.3 \\ 0 & 0 & 0 & 0 & 0.7 \\ 0 & 0 & 0 & 0 & 0 \end{bmatrix} \quad \text{where } i = 1 \text{ to } 5 \tag{7-36}$$

Suppose that now expert opinion indicates that the relative importance of each distress indicator in category 2 can be represented by the weighting fuzzy set:

$$W_{i,2} = \begin{bmatrix} 0 & 0 & 0.1 & 0.2 & 0.7 \end{bmatrix} \quad \text{where } i = 1 \text{ to } 5 \tag{7-37}$$

Referring back to Table 7-11, the first element in Equation (7-37) indicates a zero weighting to pit < 10%. Conversely, pit > 80% (the last element in Equation 73-7) is assigned a weighting of 0.7. Equations (7-35) to (7-37) can now be combined to obtain the fuzzy set for category 2, H_2. In this case, matrix multiplication gives H_2 = [0, 0, 0.02, 0.14, 0.35, 0.49, 0], which indicates the importance of category 2 towards the overall condition rating. The calculation above can be repeated for all categories and their distress indicators, which results in individual fuzzy sets H_j for each category.

Step 3. Aggregation of Categories Towards Condition Rating

Having aggregated individual distress indicators towards their respective categories, a similar calculation is conducted to aggregate categories towards condition ratings for pipes (Rajani et al., 2006). In the same way as importance weightings are assigned to individual distress indicators within a category, expert opinion is used to assign relative weightings to categories towards the condition rating. W_j denotes the relative weight of category j towards the condition rating and the aggregation process is written as (Rajani et al., 2006):

$$C = \left[W_1 ... W_j ... W_M\right] \bullet \left[H_1 ... H_j ... H_M\right]^T \tag{7-38}$$

In Equation (7-38), M is the total number of categories that influence the condition rating. W_1 to W_M are the expert opinion-based importance weightings for each category towards the condition rating. As demonstrated above H_1 to H_M are 7-element fuzzy sets (vectors) for each category and $[H_1..H_j..H_M]$ is a 7×M matrix and the '•' operator represents vector-matrix multiplication. The result C is a 7-element fuzzy set, which contains the membership values to the linguistic constants describing condition rating: *excellent, good, adequate, fair, poor, bad and failed.* For example, a resulting fuzzy set C = [0, 0, 0, 0.41, 0.59, 0, 0] would indicate membership values of 0.41 to *fair* and 0.51 to *poor*.

Fuzzy Logic-Based Approach to Modeling Deterioration

In the previous section, a documented methodology for translating observed distress indicators into a fuzzy condition rating was described (Rajani et al., 2006). In an extension to this methodology, Kleiner et al. (2006) have demonstrated how fuzzy rule-based modeling can be used to determine deterioration rate. In fuzzy rule-based modeling, the relationships between variables are represented by fuzzy *if-then* rules of the form "*If* antecedent proposition, *then* consequent proposition". *If-then* rules can be compounded with multiple antecedents and consequences. For example, a particular rule might be: *If* 'pipe age' is 'young' and 'condition' is 'poor' *then* 'deterioration rate' is 'fast'. According to Kleiner et al. (2006), the formal representation of a fuzzy rule is:

$$R_i\text{: } If\ x \text{ is } A_j,\ then\ y \text{ is } B_k; \qquad i = 1,2,\ldots,L;\ j = 1,2,\ldots,M;\ k = 1,2,\ldots,N \qquad (7\text{-}39)$$

where R_i represents the rule set (consisting of L different rules), x is the input (antecedent) linguistic variable and A_j is an antecedent linguistic constant. Equation (7-39) assumes there are M possible linguistic constants to which x can be assigned. Similarly, y is the output (consequent) linguistic variable and B_k is a consequent linguistic constant (one of N defined constants). The values of x, y, A_j and B_k are fuzzy sets defined over their respective universes of discourse. The linguistic variables A_j and B_k are predefined terms (such as 'good', 'poor' 'slow' 'fast,'). The rule set R_i and the sets A_j and B_k represent the knowledge base of the linguistic model. In the fuzzy rule-based model developed by Kleiner et al. (2006), the knowledge-base was defined by the TFNs (shown in Figure 7-11). The deterioration rate is expressed in multiples of d_0, which, as described by Kleiner et al. (2006), is the parameter used to fit the model to observed data.

Age	$q_{1,1}$	$q_{1,2}$	$q_{1,3}$
New	0	0	20
Young	0	20	40
Medium	20	40	70
Old	40	70	100
Very old	70	100	100

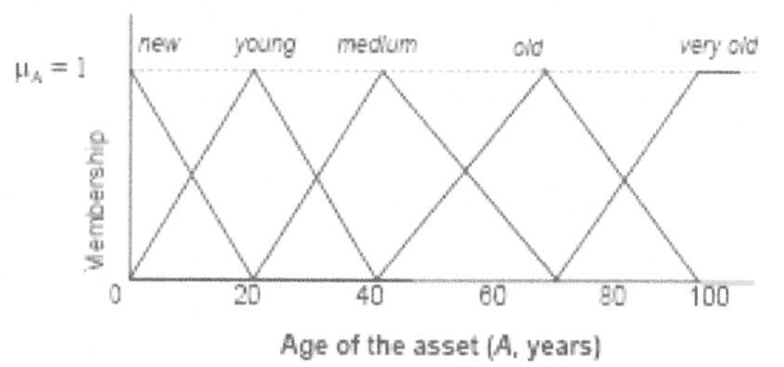

Condition state	$q_{1,1}$	$q_{1,2}$	$q_{1,3}$
Excellent	0	0	0.17
Good	0	0.17	0.33
Adequate	0.17	0.33	0.50
Fair	0.33	0.50	0.67
Poor	0.50	0.67	0.83
Bad	0.67	0.83	1
Failed	0.83	1	1

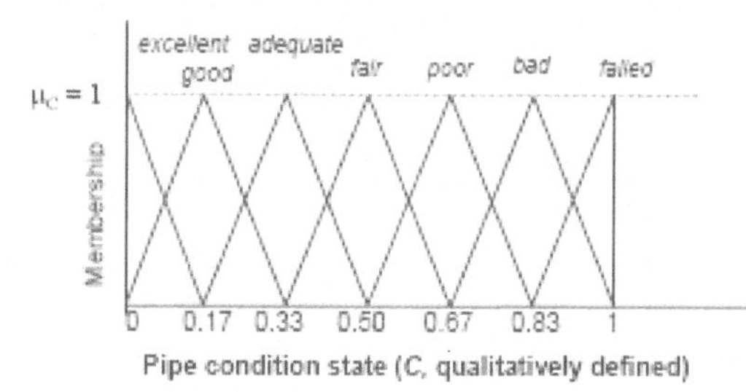

Deterioration rate	$q_{1,1}$	$q_{1,2}$	$q_{1,3}$
Very slow	0	0	d_o
Slow	0	d_o	$2d_o$
Average	d	$2d_o$	$3d_o$
Fast	$2d_o$	$3d_o$	$4d_o$
Very fast	$3d_o$	$4d_o$	$4d_o$

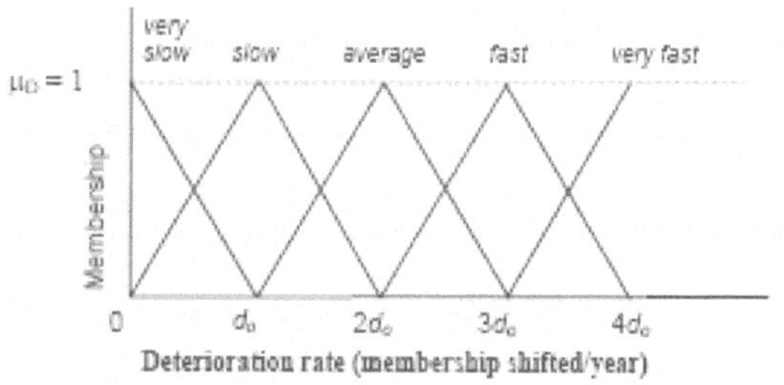

Figure 7-11. Fuzzy Sets: Age, Condition and Deterioration Rate.
(Reprinted with permission from Kleiner et al., 2006)

The fuzzy rule-set that is used in conjunction with the knowledge base to infer deterioration from age and condition is shown in Table 7-12.

Table 7-12. Fuzzy Rule Set for Inferring Deterioration Rate.
(Reprinted with permission from Kleiner et al., 2006)

R_i = *If* asset age (A) is "*A*" and pipe condition state (C) is "C" then deterioration rate (D) is "*D*" (at time = t)

Pipe condition (C)	*excellent*	*good*	*adequate*	*fair*	*poor*	*bad*	*failed*
Age (A): new	slow	average	fast	very fast	very fast	very fast	very fast
young	slow	average	fast	fast	fast	very fast	very fast
medium	very slow	slow	average	average	fast	fast	very fast
old	very slow	very slow	slow	slow	average	average	fast
very old	very slow	very slow	very slow	slow	slow	average	average

As described by Kleiner et al. (2006), the so-called 'Mamdani Inference' algorithm is used to determine the fuzzy set for deterioration rate given the fuzzy sets for pipe age and condition. While Kleiner et al. (2006) document the formal mathematical description for the Mamdani algorithm, useful descriptions are also contained in web-based Fuzzy Logic user guides (available at http://www.mathworks.com/access/helpdesk/help/pdf_doc/fuzzy/fuzzy.pdf). Kleiner et al. (2006) summarize the Mamdani inference algorithm as:

$$D_t = (A_t \text{ and } C_t) \circ R_D \tag{7-40}$$

where D_t is the 5-element fuzzy set containing membership values to linguistic descriptions of deterioration rate *very slow, slow, average, fast, very fast*; A_t is the 5-element fuzzy set for pipe age containing membership to *new, young, medium, old, very old* and C_t is the fuzzy set for pipe condition containing membership values to different states. R_D represents the fuzzy rule base and '$\circ$' is a max-min composition operator.

In essence, given that for a particular case, fuzzy sets are known for pipe age and condition, the Mamdani algorithm searches all rules in the fuzzy rule set and calculates membership values to different deterioration rates. The use of the Mamdani algorithm to infer a fuzzy deterioration rate is best illustrated through an example. Suppose that a particular pipe is 25 years old and has had its condition determined. The following sets are assumed to apply for fuzzy pipe age (A_t) and fuzzy condition (C_t):

A_t = [0, 0.75, 0.25, 0, 0] (corresponding to A = 25 years old in Figure 7-11)

C_t = [0, 0, 0.45, 0.55, 0, 0, 0] (corresponding to C = 0.42 in Figure 7-11)

It is assumed that we require the membership to "fast" deterioration rate to be calculated. Referring back to Table 7-12 it can be seen that the following seven rules in the fuzzy rule set are relevant to a "fast" deterioration rate:

1. *If* age (A_t) is "new" and condition (C_t) is "adequate" *then* deterioration rate (D_t) is "fast"
2. *If* age (A_t) is "young" and condition (C_t) is "adequate" *then* deterioration rate (D_t) is "fast"
3. *If* age (A_t) is "young" and condition (C_t) is "fair" *then* deterioration rate (D_t) is "fast"
4. *If* age (A_t) is "young" and condition (C_t) is "poor" *then* deterioration rate (D_t) is "fast"
5. *If* age (A_t) is "medium" and condition (C_t) is "poor" *then* deterioration rate (D_t) is "fast"

6. *If* age (A_t) is "medium" and condition (C_t) is "bad" *then* deterioration rate (D_t) is "fast"
7. *If* age (A_t) is "old" and condition (C_t) is "failed" *then* deterioration rate (D_t) is "fast"

The schematic diagram shown in Figure 7-12 indicates how the Mamdani algorithm would calculate the membership value to a "fast" deterioration rate for Rule 3 in this example.

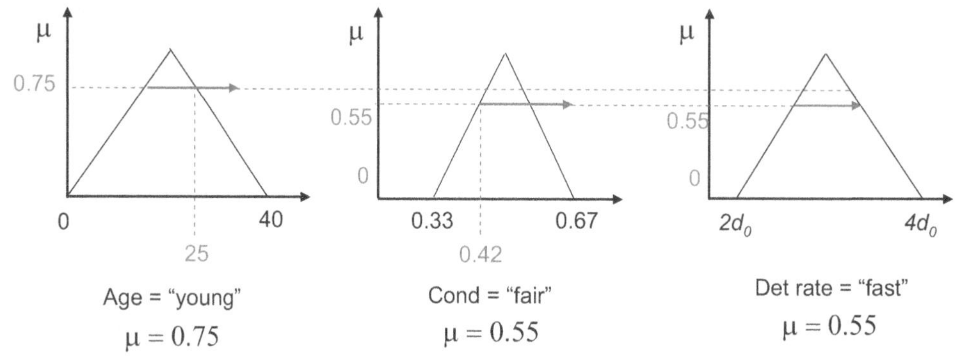

Figure 7-12. Determining Membership to "Fast" Deterioration Rate for Rule 3.

As shown, for the known fuzzy sets A_t and C_t defined in this example, the minimum membership to D_t = "fast" is 0.55 for Rule 3. The procedure in Figure 7-12 is repeated for each of the 7 rules that apply to a "fast" deterioration rate. The maximum value of membership is then assigned to that particular deterioration rate. The process is repeated to calculate membership values for all other deterioration rates in the rule set. The output from the Mamdani algorithm is therefore a fuzzy set containing membership values to the different fuzzy deterioration rates that are defined.

Once the fuzzy deterioration rate is determined, Kleiner et al. (2006) demonstrate how it can be 'defuzzified' to a crisp number (D_c) using the centroid method. Essentially, the centroid method selects the deterioration rate that divides the membership 'area' equally. The example in Figure 7-13 shows how the fuzzy deterioration rate D_t = [0, 0, 0.3, 0.7, 0] is defuzzified using the centroid method. The red line indicates the graphical representation of D_t. In this case, the crisp deterioration rate by given by the centroid method is approximately $D_c = 2.6d_0$.

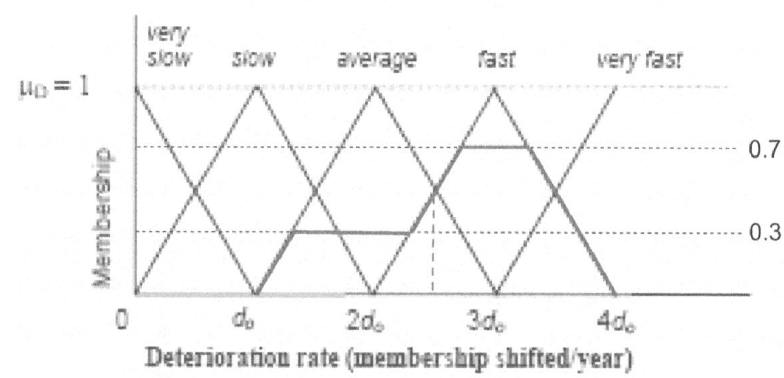

Figure 7-13. Schematic Illustration of Defuzzification.
(Reprinted with permission from Kleiner et al., 2006).

Having determined the crisp deterioration rate for a particular pipe, Kleiner et al. (2006) demonstrate how its *future* condition can be determined. If the fuzzy set for current pipe condition is known and the defuzzified deterioration rate can be determined through the process outlined above, the fuzzy condition set at any time in the future can be determined through simple matrix multiplication.

While the background theory and calculation of fuzzy deterioration rates is explained above, it should be noted that, in order to predict accurately future condition of a pipeline, some form of model calibration against real data is also required. As described by Kleiner et al. (2006), a non linear regression procedure is used for model training. In this procedure, the deterioration rate d_0 is chosen such that the sum of the squared deviations between observed and predicted membership values to different condition states is minimized. To apply this non-linear regression, the initial fuzzy set for the condition of an 'as-produced' pipe is assumed known and at least one observation fuzzy set is required. An example of the resulting condition profile from the model is shown in Figure 7-14.

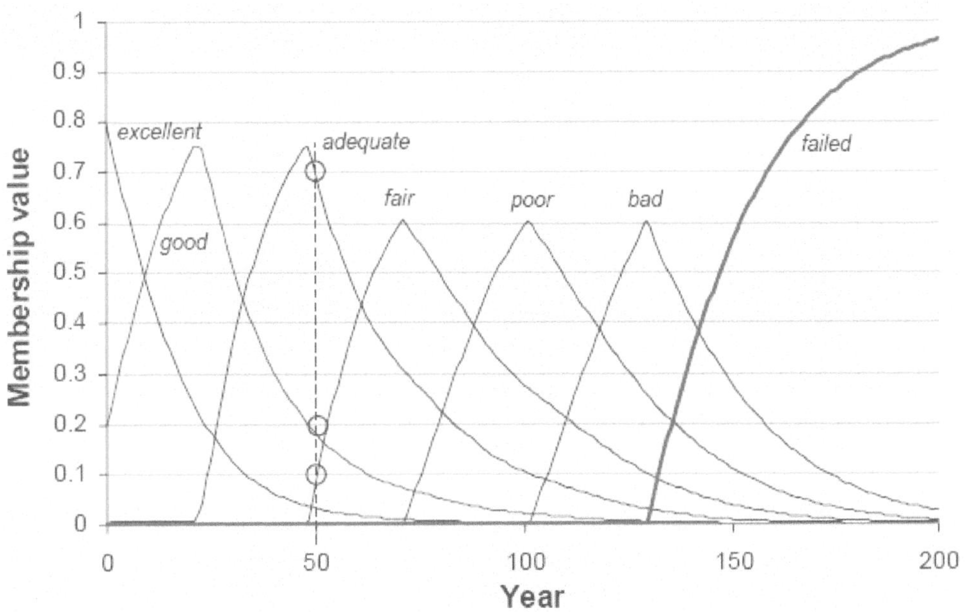

Figure 7-14. An Example Fuzzy Condition Profile.
(Reprinted with permission from Kleiner et al., 2006).

As shown, the model predictions allow the membership values to different condition states to be determined as a pipe ages. In this particular case, membership to the "failed" condition state does not activate until the pipe reaches an age of approximately 130 years.

7.15.3 Practical Considerations

As part of the artificial intelligence/soft computing modeling category, fuzzy logic-based approaches also risk being treated as "black-box" solutions. Without a thorough grounding in the background fuzzy set theory, models can be developed that are inaccurate and unrealistic. Since the application of fuzzy set theory to pipe deterioration is relatively new, significant further work is required to explore its robustness. For example, Kleiner et al. (2006) report that in one case, adding extra observed condition data to the model calibration actually degrades the accuracy of forecasting rather than improving it.

The fuzzy logic-based approach to deterioration modeling is essentially a vehicle to incorporate and handle expert opinion. As described by Rajani et al. (2006), the weighting parameters required to translate distress indicators into condition are assigned through expert opinion. As such, it is essential that an appropriate surveying technique is employed to extract this opinion. One such approach is the Delphi method, which employs a series of questionnaires designed to elicit and develop individual responses to the problem posed to a pre-selected panel of experts (Rajani et al., 2006).

7.15.4 Examples of Application

The majority of previous research on the development of fuzzy logic-based deterioration modeling for pipe assets has originated from the Buried Utilities research group in the National Research Council of Canada. For example, Rajani et al. (2006) collated relevant distress indicators for Pre-stressed Concrete Cylinder Pipe (PCCP) and Cast/Ductile Iron pipe. Rajani et al. (2006) also developed the fuzzy synthetic evaluation method and demonstrated how a fuzzy condition set could be determined for PCCP. The application of this method has been demonstrated for a 96-inch diameter PCCP that was installed in 1978 and inspected again in 1997 and 1999. The distress indicators that were used to determine the fuzzy condition set at each inspection are shown in Table 7-13. These distress indicators were combined with importance weightings obtained from expert opinion.

Following this work, Kleiner et al. (2006) developed and calibrated a fuzzy rule-based deterioration model and applied it to the 96-inch diameter PCCP. The model was trained using the inspection data set from 1997 and validated against the inspection data from 1999. Reasonable agreement was obtained for validation and the model indicated that the pipe entered the "failed" state at an age of 65 years.

Table 7-13. Distress Indicators Required for the Fuzzy Condition Set of PCCP Pipe.
(Reprinted with permission from Rajani et al., 2006)

Category (Level 1)	*(j)*	Distress indicator (Level 2)	*(ij)*	Comment
Mortar coating	1	Spalling	1,1	Spalling is often a first indicator of corrosion. Large areas may indicate that corrosion is taking place over a significant surface area of pipe exterior.
		Crack type	2,1	Circumferential cracks indicate some type of longitudinal movement has taken place. Longitudinal cracks occur due to low hoop resistance (wire breaks?).
		Crack width	3,1	Crack width is another indicator of severity of spalling. Large widths mean that spalling is imminent.
		Crack density (frequency)	4,1	Closer crack spacing usually means the pipe is under higher stress.
		Coloration	5,1	Signs of color/stains on concrete exterior indicate that corrosion is taking place. Often stains are precursors to spalling i.e., corrosion products have built up.
Pre-stressed wire	2	Wire breaks	1,2	As the number of wire breaks increase the factor of safety decreases and eventually leads to pipe failure.
Concrete core	3	Delamination	1,3	Delamination occurs when there is poor bonding between concrete/wire or steel/steel cylinder. This can also occur when prestressing is lost due to wire breaks.
		Crack type	2,3	Circumferential cracks indicate some type of longitudinal movement has taken place. Longitudinal cracks occur due to low hoop resistance (wire breaks?).
		Crack width	3,3	Crack width is another indicator of severity of spalling. Large widths mean that spaling is imminent.
		Crack density (frequency)	4,3	Closer crack spacing usually means the pipe is under higher stress.
		Hammer tapping sounding	5,3	Hammer tapping sounds can indicate delamination. It can be simple as tapping a hammer or using 'pulse echo' method.
		Hollow area	6,3	Aerial extent of hollow sound heard can give an idea of the seriousness of the lamination (in comparison to pipe surface area).
Pipe geometry	4	Out of roundness	1,4	Out-of-roundness is another indicator of wire loss that may not be evident from concrete spalling or presence of corrosion products, etc.
Joint	5	Change in alignment	1,5	Changes in joint alignment indicate pipe susceptible to ground movement. Eventually it can lead to weld failures and hence joint failure.
		Joint (internal) displacement	2,5	Joints can displace without undergoing joint misalignment and hence also an indicator of other forces at play.
		Joint diaper crack size	3,5	Crack of external diaper can give an idea of joint quality.

7.16 Technique Application as a Function of Asset Management Capacity

It is interesting to consider how a water authority might develop its approaches to the analysis of remaining asset life as it develops its asset management capacity. While this can be anticipated to be a path dependent process to a degree, it is still informative to illustrate conceptually how this development might occur. To this end, Table 7-14 details a view of asset management development, along with the approaches to assessment and analysis of remaining life that could be used at each stage. It is assumed that a more sophisticated approach to asset management gives rise to more data availability, and consequently the possibility of more approaches to modeling remaining asset life.

In reality, the selection of tools will depend upon the level of asset management sophistication, as well as the capacity of the tools themselves. With this in mind, Table 7-15 attempts to map tools outlined in this chapter on to the following selection factors:

1. ***Minimum asset management sophistication***: Do the requirements of the modeling technique imply a minimum level of asset management sophistication needs to be attained before they can be used?
2. ***Data requirements***: Are there high level of data requirements?
3. ***Inclusion of expert opinion***: Does the approach allow expert opinion to be utilized?
4. ***Treatment of uncertainty***: Does the technique explicitly consider uncertainty?
5. ***Treatment of economic risk factors***: Does the technique consider economic risk factors, or does it simply consider physical asset life?
6. ***Blackbox***: Does the technique provide a framework within which to understand the problem or is it more of a blackbox approach, providing answers with no specific insights into the mechanisms of the decision-making?
7. ***Packaged***: Is the technique packaged and available off the shelf?
8. ***Population or single asset***: Does the technique apply to a single asset, population or both?
9. ***Scenario testing***: Does the technique provide the capacity to test different scenarios and thus understand the results of different strategies?

The level of asset management sophistication attained does not necessarily relate to the size of company, since relatively small companies can develop sophisticated approaches for some asset classes. In fact, the level of sophistication that adds value to a particular water authority will depend on the nature of its asset stock, its business environment (in particular, the requirements of stakeholders), resource constraints and technical capacity. For example, the research team has previously worked with a company that operates an extensive pipe network but only one treatment work. This company had a relatively sophisticated approach to the management of its pipe network, but a relatively unsophisticated approach to its management of treatment work assets, since this was all that was considered necessary to meet its asset management needs. Interestingly, a change in requirements from a regulator eventually drove the company in question to adopt a more sophisticated approach to management of its treatment work assets.

It is also noteworthy that water authorities need to focus on the stages of asset management that are most relevant to their business. For example, companies with relatively new and still expanding asset stocks (e.g., due to new developments) need to develop good practice in the early life cycle stages of asset management, whereas those that are managing aging asset stocks need to focus more on rehabilitation planning and delivery. The latter would require more sophistication in the evaluation and prediction of remaining asset life than the former.

7.17 Research Challenges: Modeling Remaining Asset Life

From the information presented in this chapter, the research challenge associated with assessing remaining life can be summarized thus:

1. There is a need for processes which align the different available techniques to different asset types and scenarios – i.e., industry guidance on what is the best approach to adopt and why.
2. For statistical models, further work is required to identify the optimum data requirements for model calibration.
3. For physical probabilistic models, technologies to support condition assessment to quantify model variables for all asset types need to be developed. Protocols on the use of probabilistic models would also be useful.
4. The use of soft-computing/artificial intelligence-based approaches for asset life time prediction should be explored further. Specifically, methods to incorporate useful, and readily available engineering expertise into the model development should be pursued.
5. For all modeling approaches, benchmarking to demonstrate their validity is essential.
6. A road map is needed that demonstrates the research and development required for water authorities to increase their capacity in analysis of remaining asset life, as well as to provide guidance so that water authorities can select an appropriate level of sophistication, which takes into account their business environment and operational constraints.

Table 7-14. Conceptual Model of Tool Use According to Development of Asset Management.

Stage of Asset Management Development	Characteristics of Asset Management Function	Assessment and Analysis of Remaining Life
No formal systems	No specific asset management function, much of the maintenance and planning is undertaken reactively and is not optimized to any degree.	– Where remaining life estimates are used, expert opinion driven approaches are adopted in conjunction with assumed asset lives (e.g., based on manufacturer's estimates) – Condition assessment and risk assessment may be applied to specific assets, but their use is ad hoc rather than systematic – Deterministic models can be applied as part of these assessments – Analysis of interventions tends to focus on financial issues (internal costs)
Initial stages; understanding what assets are owned and what state they are in	Developed inventory of assets with associated data, but still generally reactive approach to asset maintenance and planning, but with some proactive management of specific 'key assets'; basic asset management plans for main asset types exist.	– Expert opinion driven approaches used in conjunction with assumed asset lives – Condition and performance grading techniques used as the basis for long term planning – Condition assessment and risk assessment may be applied to specific assets, but their use is ad hoc rather than systematic – Deterministic models can be applied as part of these assessments – Cohort based approaches possible for pipes (but expert opinion used to develop survival curves or similar) – Probabilistic models possible but likely to be viewed as overly sophisticated – Fuzzy logic approaches could be used, but this is unlikely – Analysis of interventions tends to focus on financial issues (internal costs)
Basic asset management function	Basic maintenance management system now in place, some additional supporting asset management functions with respect medium to long term planning. General focus on management of condition and performance, but some ability to optimize maintenance and planning on the basis of cost.	– Expert opinion driven approaches still used in conjunction with assumed asset lives, but now supplemented through analysis of asset-specific data – Condition and performance grading techniques still used within planning, but the scope of the analysis is now broadened to take into consideration risk grades – Condition assessment and risk assessment of specific assets or asset types used in response to specific drivers – Deterministic models can be applied as part of these assessments – Cohort based approaches possible for pipes (survival curves and similar may be developed through expert opinion and/or analysis of works orders) – Basic statistical analysis of failure data possible, especially for M&E assets. – Use of probabilistic models possible, but perhaps still seen as overly complex. – Some optimization of maintenance costs through appropriate strategy (preventive, reactive, proactive as needed) – Fuzzy logic approaches could be used, but this is unlikely – Analysis of interventions tends to focus on financial issues (internal costs)

Stage of Asset Management Development	Characteristics of Asset Management Function	Assessment and Analysis of Remaining Life
Advanced asset management function	Advanced maintenance management with GIS capability, well developed business processes for tactical and strategic asset management, well defined KPIs and focus on risk in planning and maintenance. Able to optimize spending according to service levels and risk.	– Expert opinion driven approaches still used in conjunction with assumed asset lives, but now embedded in business rules and analytical approaches – Condition and risk grading techniques used for prioritization processes only, rather than planning – Condition assessment and risk assessment of specific assets or asset types used within a systematic framework – Deterministic models can be applied as part of these assessments – Cohort based and asset-level approaches used for long term planning (survival curves and similar developed through analysis of failure and other supplementary data) – Refined statistical analysis of failure data and service levels possible for all assets with appropriate levels of failure data – Use of probabilistic models possible to interpret inspection data and provide better insights into economic life – Fuzzy logic approaches could be used, as could ANN – Optimization of maintenance costs through adoption of techniques such as Reliability Centered Maintenance (RCM) – Analysis of interventions still tends to be driven by financial considerations, but more emphasis is given to triple bottom line issues
Whole of business strategy	Fully integrated data systems and whole of business approach to asset management with supporting functions. Focus on service and risk extended to consider broader definitions of 'added value' considering triple bottom line outputs; KPIs refined to support this approach. Able to optimize spending to meet broader business values, as well as cost, etc.	– Expert opinion driven approaches only used as a feed in to appropriate analytical techniques – Condition and risk grading techniques used for prioritization processes only – Condition assessment and risk assessment of specific assets or asset types used within a systematic business performance framework – Cohort based and asset-level approaches used for long term planning (survival curves and similar developed through analysis of failure and other supplementary data) – Refined statistical analysis of failure data and service levels undertaken for all assets and embedded into decision support processes – Use of probabilistic models undertaken to understand condition assessment data, and provides better insights into economic life – Fuzzy logic approaches could be used, as could ANN – ANN techniques are used to interpret extensive asset related data – Optimization of maintenance costs through adoption of techniques such as Reliability Centered Maintenance (RCM) – Full economic analysis undertaken of all strategies and interventions, considering triple bottom line elements (including intangible costs and externalities)

Table 7-15. Mapping Tools on to Asset Management and Other Factors.

	Deterministic Approaches	Statistical Models – Failure Event Data	Statistical Models – Service Lifetime	Statistical Models – Cohort Survival	Statistical Methods – Ordinal Regression
Minimum AM sophistication	None	Basic	Basic	Initial stages	Initial stages
Data requirements	Low	Asset and failure data	Asset and failure data	Asset and condition data	Asset and condition data
Inclusion of expert opinion	None	Only in model construction	No	Only in model construction	Only in model construction
Treatment of economic risk factors	None	Yes, when packaged	Possible	Possible	No
Blackbox	No	No	No	No	No
Packaged	No	Packages available to facilitate analysis; packages for specific asset types available	Packages available to facilitate analysis	Packages available to facilitate analysis; packages for specific asset types available	Packages available to facilitate analysis
Population or single asset	Single asset	Population	Population	Population	Population
Scenario testing	Not explicitly	Yes when packaged	No	Yes when packaged	No

	Statistical Models – Markov Chain	**Statistical Models – Bayesian Approaches**	**Physical Probabilistic Models**	**Structural Reliability-Based Methods**	**Artificial Neural Networks**	**Fuzzy Logic**
Minimum AM sophistication	Initial stages	Basic	Any	Any	Advanced	Initial stages
Data requirements	Asset and condition data	Asset and failure data	Asset data	Asset data	Asset and failure data	Asset data
Inclusion of expert opinion	Only in model construction	Only in model construction	Only in model construction	Only in model construction	No	Yes
Treatment of economic risk factors	No	No	Possible	Possible	No	Possible
Blackbox	No	No	No	No	Yes	Could be
Packaged	Packages available to facilitate analysis; packages for specific asset types available	Packages available to facilitate analysis	Packages available to facilitate analysis	Packages available to facilitate analysis	Packages available to facilitate analysis	Packages available to facilitate analysis
Population or single asset	Population, but potentially asset level	Population	Singe asset	Singe asset (population also possible)	Population	Population
Scenario testing	No	No	Possible	No	No	No

CHAPTER 8.0

RECOMMENDATIONS

Recommendations for further research and development are identified at the end of each chapter, and, for completeness, the main ones are summarized below.

Standard Terminology. While the treatment given in this report represents a starting point, there remains a need for industry standardization of terms and definitions, as well as a consistent way to account for obsolescence in the consideration of asset management and remaining life. In addition, work should be undertaken to develop further the list of factors that influence asset life for key asset types, and to identify the most significant of these factors (using analysis of data, expert opinion, industry reviews, etc.).

Incorporating Risk Information. Several recommendations relate to the development of decision-support protocols. In a broad context, it would be useful to develop a protocol to specify what approaches to assessment of risk are suitable for different asset management needs, based on the relative cost of alternative assessment procedures, and the uncertainty and risk inherent in the decision making process. A first step for water authorities beginning to develop their asset management capabilities would be to standardize risk matrices and their application in asset management decision making. Further, there needs to be a determination of how best to contextualize assessments of condition in terms of risk, and to do this in a way that ensures interventions can be appropriately timed to meet asset management objectives.

The use of life cycle costing techniques should also be developed, including the collection of cost data at different times in the asset life cycle, the specification of analysis approaches for a range of asset classes, and dealing with associated uncertainty in decision-making. Standardized approaches are needed to:

1. consider external costs and externalities in asset management decision-making, including methods to place a monetary value on these costs;
2. assess the consequence of risk, taking into account the probability-side of consequence to ensure that decision-making is based on realistic assessments of business risk exposure; and
3. assess the relative importance of assets (based on consequence factors), and guidance for use, underpinned by the utilization of reliability approaches to determine failure modes for key asset types.

Spatial and Temporal Resolution. Analysis of remaining asset life also needs to incorporate processes at different levels of spatial and temporal resolution; i.e. to map approaches onto the processes used in SAM, TAM and operational risk management. The relative benefits of using more sophisticated approaches for calculating remaining life should be evaluated in comparison to the use of book-life and other simplistic approaches. Deterioration modeling should also be extended to consider service level deterioration, and not just physical asset deterioration.

The Influence of Rehabilitation Options on Risk. Research is also needed to determine (or at least constrain) the change in failure probability that can be expected from different rehabilitation options, so that this change can be used in cost-benefit analysis of intervention options. Finally, research is required to show how supply, demand, quality, and capital maintenance issues can be simultaneously integrated into the assessment of remaining asset life using a range of techniques.

Consistent Grading Schemes. For operational procedures such as condition assessment and monitoring, standard guidance is needed on what constitutes a significant defect for a range of asset types in different operational contexts. In addition, protocols are needed for qualitatively and quickly "screening" equipment or structures to identify those that are approaching the end of their life. Since much data collected in relation to this is qualitative, subjective or descriptive, protocols should be developed that ensure that assessments are consistent and free of bias. Common condition, performance and risk grading procedures and specifications should be developed across the sector and applied consistently over time. There is also a need to specify protocols for the use of these grades in asset management.

Modeling of Remaining Asset Life. Recommendations for modeling of remaining asset life include development of protocols that align the different available techniques to different asset types and scenarios. In particular, modeling research should focus on:

1. Identifying the optimum data requirements for calibration of statistical models;
2. For physical probabilistic models, link condition assessment processes to the modeling technique for all pipe materials, as well as for the development of protocols and tools to aid model use;
3. Explore the use of soft-computing/artificial intelligence-based approaches for asset life time prediction, as these models appear to fill some gaps left by other approaches, especially for more subjective data;
4. Pursue methods to incorporate useful and readily available engineering expertise into model development; and
5. Benchmark model predictions against real world data to demonstrate the validity of the various modeling approaches.

Selecting an Asset Management Approach. The progression of asset management development discussed in Chapter 7.0 provides one view of how water authorities could align the use of different tools and techniques with their level of asset management sophistication. Nevertheless, a key goal of the overall WERF research program is to produce a 'road map' that outlines the research needed to facilitate water authorities to select asset management tools that meet the needs of their business environment within operational constraints.

APPENDIX A

A Taxonomy of Remaining Asset Life

Various definitions relating to asset life and end of life criteria are included in the taxonomy given in Chapter 2.0. This Appendix provides a more detailed consideration of the various definitions included in the taxonomy.

Design, Service Life, and Operational Life

Amongst other definitions, the IPWEA (2006) refers to the concept of asset ***design life***, which is the period of time over which the asset is expected to be available for use and able to provide the required level of service at an acceptable cost and level of risk. Design life is inherently associated with the expected life of the asset at the start of the asset's life cycle.

The ***service life*** of an asset is taken to be the length of time over which the asset is actually available for use and able to provide the required level of service at an acceptable cost and level of risk. An asset may be operated past the point where it should be replaced for financial, risk or service considerations, thus the asset's ***operational life*** may exceed the 'service life', as defined above.

End of Physical Life

At a fundamental level, remaining asset life is related to structural integrity of an asset, in that once structural integrity has passed beyond an acceptable level, such that the asset can no longer provide the required functionality and can not be repaired, it can be considered to be at the end of its ***physical life***. Such assets will need a significant level of investment (in proportion to the cost of the asset) to bring back to the state where function can be provided or will need to be replaced. For example, a blocked sewer is not at the end of its life since it can be jetted, whereas a collapsed sewer, with high levels of deterioration in non-collapsed sections, would be considered at the end of its life.

In summary, the end of physical life occurs when the asset is structurally derelict and/or non-functional and repair (or non-structural rehabilitation in the case of sewers) is no longer a technically viable option.

End of Economic Life

Asset renewal is inherently linked to the economics factors that underlie the replacement decision. In turn, this means that the concept of remaining asset life is intimately linked with economic factors. The question is frequently not whether the asset is at the end of its physical life (i.e., is structurally derelict, see above), but whether it is more economic to retain or substantially rehabilitated or replace it. This can be influenced by a range of factors. For example, Baur and Hertz (2002), note the life of a sewer is not determined just by deterioration, but also by unit costs of repair and rehabilitation work, and by changes to technical specifications/requirements and service standards. Similar assertions can be made for other asset categories.

When considered in this light, an asset will thus have an optimal time of renewal, which may vary as technical and commercial innovations occur. The period over which it is more economic to retain an asset can be referred to as the ***economic life***, and the end of economic life can be taken to be the point in time where renewal is the most beneficial option available.

In summary, the end of the economic life for an asset can be defined as the point in time when the present value of the future operation, maintenance and risk costs exceeds the current costs of replacing or rehabilitating the asset.

End of Service Life

The service life of an asset is defined as the period of time over which the asset is actually available for use and able to provide the required level of service at an acceptable risk; e.g., without unforeseen costs of disruption for maintenance and repair. End of service life implies that the asset is no longer able to provide the required level of service and/or there are unforeseen maintenance costs and/or repairs.

In some cases, end of service life is driven solely by service provision considerations, rather than financial ones. A good example is a sewer that blocks repeatedly. From a purely financial perspective, it may be more advantageous to keep on clearing the blockage. However, minimum service provision considerations (in this case, expressed in terms of acceptable blockage rates) can mean such an option is not, in practice, available. As discussed in Chapter 3.0, it is also sometimes necessary to consider service risk: i.e. assets may be replaced because they pose a risk to future service provision, rather than because of a poor service history.

In summary, the end of service life occurs when the asset is no longer able to provide the required level of service at an acceptable level of risk.

End of Financial Life

Every asset owned by a water authority is subject to deterioration over some time scale (whether months to decades) resulting in a reduction in future service potential. This consumption of service potential is recognized in financial reports as a loss of capital value; this loss of value is termed depreciation.

In financial accounting, depreciation is a method for attributing the cost of an asset across its life. In practice, the capital value of the asset is spread over the period in which it is expected to provide service. The capital value of the asset can be expressed in historical terms (i.e. the original cost of acquiring the asset in question), or in 'modern equivalent terms' (i.e. the cost of replacing the asset with a modern equivalent asset; the modern equivalent could be cheaper or more expensive due to technical innovation and other commercial factors).

It is a common misconception that depreciation is charged to provide for replacement of assets in a company's finances. In reality, depreciation is used to apportion the cost of the asset over its expected lifespan, so that the cost of production can be appropriately reflected in the price of the product. For example, if an asset was bought for $1,000,000 and it has an expected life span of 20 years and can produce the same amount of product each year, rather than charge all the purchase cost in the first year, the asset would be depreciated at $50,000 per year ($1,000,000/20 years). Hence, a cost of $50,000/year would be charged in financial reports over the life time of the asset. This allows the cost of the asset to be apportioned against its yearly production over the time it is expected to be in operation.

A report by LGA (2001), gives a useful summary of some traditional depreciation methodologies. These methods are formula-based and assume that an asset has a finite life:

- ***Straight-line depreciation***: this approach assumes a uniform depreciation expense over the life of the asset. This method is used where the asset is expected to be used on a uniform basis over its estimated life. The formula for a given asset is:

  (gross cost) - (estimated residual or salvage value)/(estimated life)

  For example, a truck purchased for $120,000, which is expected to be sold for $20,000 after five years service would have an annual depreciation of ($120,000 - $20,000)/5 = $20,000.

- ***Reducing balance depreciation***: where a fixed percentage is applied to give higher depreciation early in the life of the asset. This method is applied where most use of the asset is likely to be early in its life and decrease over time, or where there is a strong likelihood of technical or commercial obsolescence (see next section) early in the life of the asset. For example, a computer system purchased for $300,000, which is likely to have a life of three years, but whose value is quickly reduced by technical obsolescence, may have a percentage of 60% applied as a depreciation rate as follows:

Year 1	$300,000 x 0.6 = $180,000	(balance $120,000)
Year 2	$120,000 x 0.6 = $72,000	(balance $48,000)
Year 3	$ 48,000 x 0.6 = $28,800	(balance $19,200)

- ***Usage basis***: applied where the deterioration (i.e. wear and tear) of an asset is directly related to its usage. The usage could be production hours, number of items manufactured, distance traveled or some other usage criterion. For example, pumping equipment may be depreciated on an hours run basis based on a life of, say 60,000 hours. The capital value of the pump would then be consumed according to the hours run expressed as a proportion of the expected life. So for the example pump with an expected life of 60,000 hours, after 30,000 hours operation, half the capital value would have been consumed (the asset is depreciated by 30,000/60,000=0.5).

From the perspective of remaining asset life, a key issue is that traditional approaches to depreciation require an estimation of life for the asset category in question. This assumed life is often referred to as the 'book life' of the asset. The service/operational life of an asset can vary according to various factors and it is common for the actual life to exceed the book life. When this occurs, the asset value is fully depreciated in financial terms, and the asset can thus be considered to be at the ***end of its financial life***. However, since it is still providing service, capital investment can be deferred into the future. For this reason, depreciation does not necessarily have a real meaning in an asset management context; i.e. ***the fact that an asset is fully depreciated does not mean that it is at the end of its useful life***.

To avoid this issue, the concept of ***condition-based depreciation*** can be applied. In this approach, the level of depreciation applied to an asset explicitly reflects any assessments of the remaining life of that asset, still having regard to such factors as asset usage and the rate of technical and commercial obsolescence (IPWEA, 2006). Condition-based depreciation thus recognizes the use made of an asset over a particular time period based on the change in the condition of the asset and other factors, rather than from the perspective of an assumed life.

In summary, end of financial life occurs when an asset is fully depreciated and thus no longer has any 'book value'.

Obsolescence

Another factor that can influence when an asset is considered at the end of its life is obsolescence. Assets can become obsolete as a result of developments in technology. In some cases, this can lead to the situation where it is cost-effective to replace the asset well before its design life has been reached. The asset may still be functioning, but improvements in performance available with the latest generation of the equipment can make it desirable or essential to upgrade. A common example is information technology (IT) based assets, where the current rapid rate of technical development can necessitate replacement long before assets actually cease to function. Another example is the case of older pumps that are still functioning, but energy inefficient in comparison to designs that are more modern.

Equipment can also become obsolete due to changes in regulations covering its use, including changes in health and safety regulations that cannot be met by retrofitting existing equipment in a cost effective way. For example, changes in electrical regulations can lead to equipment being replaced because they cannot be modified to meet the new regulations. Environmental regulations can similarly drive replacement of assets that can not meet new standards.

From the above descriptions, it is clear that obsolescence can be driven by a number of issues that are not directly related to the state of the asset itself. As such, various types of obsolescence can be identified; for example:

- ***Technological obsolescence***; occurs when assets have been superseded by assets using more cost-effective methods of providing the same level of utility; e.g., more efficient and less expensive pumps (IPWEA, 2006). Technical obsolescence can also be considered to occur when spare parts or technical skills for supporting/maintaining an asset are no longer available.
- ***Regulatory obsolescence***; this type of obsolescence is brought about by modifications to regulatory requirements;
- ***Functional obsolescence***: obsolescence arising when existing equipment cannot be brought into line with new (non-regulatory) operating requirements.

Obsolescence is most likely to occur with long-life equipment or equipment that is subject to rapid technological development such as computers, software, monitoring instrumentation and control systems (i.e. ICA assets, see Chapter 1.0). In the later case, production lifetimes are short and support for the equipment may be withdrawn by the manufacturer.

The decision to replace obsolete assets will be driven by a range of factors including the availability of spares, the requirement to integrate new equipment with old, the economic benefits of updating (e.g., cost/performance) and any requirement to comply with more stringent operational or safety standards.

Previous experience of replacement due to obsolescence will provide some measure of the likely impact of obsolescence on asset life. Such considerations can in turn influence the design of assets. For example, regulatory obsolescence may mean it is pragmatic to adopt lower design standards (designs with shorter expected physical lives) or increase rates of depreciation (the asset is written off over a shorter period).

In summary, obsolescence occurs when technical innovations or regulatory requirements drive the replacement of an existing asset that would otherwise being considered to be in an acceptable state.

GLOSSARY

Assessment: Putting the inspection and other data into context and making a judgment as to the condition of the system or asset.

Asset: A physical component of a facility that has value, enables services to be provided, and has an economic life of greater than 12 months.

Asset Failure: When an asset is no longer providing the service or function for which it is intended.

Asset Management: The combination of management, financial, economic, engineering and other practices applied to assets with the objective of providing the required level of service to customers and the environment at acceptable levels of risk and in the most efficient manner.

Asset Observation: All information that relates to assets, their environment, and operation.

Inspection: To collect asset observations relating to condition and performance of a given asset.

Intervention: To undertake capital or operational works to repair, replace or renovate an asset.

Levels of Service: Defined measures of performance or benefit as received by the community and environment. Service levels usually relate to quality, quantity, reliability, responsiveness, environmental acceptability, and cost.

Maintenance: All actions necessary for retaining an asset as near as practicable to its original condition, excluding renewal.

Monitoring: The repeated observations over time of some aspect of an asset especially to detect and give warning of change.

Performance: A measure of a system or asset's ability to function in accordance with expected requirements or defined targets given the local operating context and intended purpose.

Rehabilitation: Works to rebuild or replace parts or components of an asset, to restore it to a required functional condition, which may incorporate some modification.

Renewal: Works to rehabilitate or replace existing assets with assets of equivalent capacity or performance capability.

Replacement: Provide a new asset in place of an existing asset that has reached the end of its life, so as to provide a similar function or level of service.

Risk: A measure of the combination of the consequence of an event occurring and the probability or likelihood of the event occurring. Often defined mathematically as the product of the probability and consequence of failure.

Structural Condition: The state of an asset at a particular time as it pertains to the degree of physical deterioration and/or propensity for failure.

Technique: A defined procedure or methodology used to accomplish a specific activity or task.

Tool: An instrument, device or software package used to collect or interpret asset observations.

REFERENCES

Abaza, K.A., Ashur, S.A., and Al-Khatib, I.A. (2004) Integrated Pavement Management System with a Markovian Prediction Model, *Journal of Transportation Engineering*, 130(1), 24-33.

ABS (2003) *Guide for Surveys Using Risk-Based Inspection for the Offshore Industry*, American Bureau of Shipping, December 2003.

ABS (2004) *Guidance Notes on Reliability-Centered Maintenance*, American Bureau of Shipping, July 2004.

ACI 318 (1999) *Building Code Requirements for Reinforced Concrete*, ACI, Farmington Hills.

Achim, D., Ghotb, F., and McManus, K.J. (2007) Prediction of Water Pipe Asset Life Using Neural Networks, *Journal of Infrastructure System*, 13(1), 26-30.

Ahammed, M. and Melchers, R.E. (1997) Probabilistic Analysis of Underground Pipelines Subject to Combined Stress and Corrosion, *Engng. Struct.,* 19(12), 988-994.

Anderson, M. and Civiti, J. (2002) Applying a Risk Management Model to Develop a 20-Year Capital Works Program for Water and Sewer Pipelines, *CSIRO/AWA 2nd Annual Seminar on Asset Management of Pipelines: Risk Based Performance of Urban Water Pipelines*, 26th November 2002, CSIRO, Highett, Victoria, Australia.

Ariaratnam, T. S., Assaly, E.A., and Yuqing, Y. (2001) Assessment of Infrastructure Inspection Needs Using Logistic Models, *Journal of Infrastructure Systems*, ASCE, 7(4), 66-72.

Arunraj, N.S. and Maiti, J. (2007) Risk-based Maintenance – Techniques and Applications, *Journal of Hazardous Materials*, 142, 653–661.

AwwaRF (2001) *Water Treatment Plant Infrastructure Asset Management: Users Manual,* prepared by L. Elliot et al., AWWA Research Foundation, Denver, CO.

AwwaRf (2005a) *Long Term Performance of PVC Pipes*, prepared by Burn, S., Davis, P., Schiller, T., Tiganis, B., Tjandraatmadja G., Cardy, M., Gould S., Sadler, P. and Whittle, A., AwwaRF Project No. 2879, Denver, CO.

AwwaRF (2005b) *Risk Management of Large-Diameter Water Transmission Mains*, prepared by Kleiner, Y., Rajani, B., and Sadiq, R., AWWA Research Foundation and American Water Works Association, Denver, CO.

Baik, H. S., Jeong, H.S. and Abraham, D. M. (2006) Estimating Transition Probabilities in Markov Chain-Based Deterioration Models for Management of Wastewater Systems, *Journal of Water Resources Planning and Management*, ASCE, 132(1), 15-24.

Baur, R. and Herz, R. (2002) Selective inspection planning with ageing forecast for sewer types. *Journal of Water Science and Technology*, 46(4-7), 389-396.

Basu, A.P. and Rigdon, S.E. (2000) *Statistical Methods for the Reliability of Repairable Systems*, Wiley series in Probability and Statistics, New York: Wiley & Sons, Inc.

Bishop, C. (1995) *Neural Networks for Pattern Recognition*, Oxford University Press, New York.

Boussabaine, A. and Kirkham, R. (2004) *Whole Life-Cycle Costing: Risk and Risk Response*, Blackwell Publishing, Oxford, UK.

Buckland, P. (2000) Risk Based Economic Life Cycle Management of Infrastructure Assets and Its Regulatory Implications, *In Proc. International Conference of Maintenance Societies*, Wollongong, Australia, May 2000.

Buckland, P. and Hastings, N. (2001) The Replacement Decision for "Linear Assets", *International Conference of Maintenance Societies – ICOMS 2001*, Melbourne, Australia - May 2001.

van Bueren, M. and Hatton MacDonald, D.H. (2004) Addressing Water-Related Externalities: Issues for Consideration, *Paper presented at a Water Policy Workshop convened by the Australian Agricultural and Resource Economics Society*, 10th February 2004, Melbourne.

Burn, S., Marlow, D., Moglia, M., and Buckland, P. (2007) Asset Management for Urban Infrastructure, *Water Asset Management International*, 3(2), 12-18.

BSI (2008) *PAS 55 Part 1: Specification for the Optimized Management of Physical Infrastructure Assets*, BSI, UK, 2008.

Cheung, M.S. and Kyle, B.R. (1996) Service Life Prediction of Concrete Structures by Reliability Analysis, *Construction and Building Materials*, 10(1), 45 -55.

Clark, R.M., Stafford, C.L., and Goodrich, J.A. (1982) Water Distribution Systems: A Spatial and Cost Evaluation, *J. Water Resources Planning and Management Division*, ASCE, 108(3), 243-256.

Constantine, A.G., and Darroch, J. (1993) *Pipeline Reliability, Stochastic Models in Engineering Technology & Management, Editors - S. Osaki and D.N.P Murthy*, Technology Management Centre, The University of Queensland, Australia, 86-95.

Crowder M.J., Kimber, A.C., Smith A.C., and Sweeting, A.J. (1991) *Statistical Analysis of Reliability Data*, Boca Raton, Florida, Chapman & Hall/CRC.

D'Agostino, R. B. and Stephens, M. A. (1986) *Goodness-of-Fit Techniques*, Marcel Dekker, Inc., New York.

Datla, S.V. and Pandey, M.D. (2006) Estimation of Life Expectancy of Wood Poles in Electrical Distribution Networks, *Structural Safety*, 28, 304-319.

Davis, P., De Silva, D., Marlow, D., Moglia, M, Gould, S. and Burn, S. (2008) Service Life Prediction and Scheduling Interventions in Asbestos Cement Pipelines, *Journal of Water Supply: Research and Technology – AQUA*, 57(4), 239-252.

Davis, P. and Marlow, D. (2008) Quantifying Economic Lifetime for Asset Management of Large Diameter Pipelines, *Journal of the AWWA,* 100(7), July 2008.

Davis, P., Burn, S., Moglia, M., and Gould, S. (2007) A Physical Probabilistic Model to Predict Failure Rates in Buried PVC Pipelines, *Reliability Engineering and System Safety*, 92, 1258-1266.

Davies, J.P., Clarke, B.A., Whiter, J.T., and Cunningham, R.J. (2001a) Factors Influencing the Structural Deterioration and Collapse of Rigid Sewer Pipes, *Urban Water Volume*: 3 (1-2), March - June, 2001, 73-89.

Davies, J.P., Clarke, B.A., and Whiter, J.T. (2001b) The Structural Condition of Rigid Sewer Pipes: A Statistical Investigation, *Urban Water*, 3, 277-286.

Deb, A.K., Snyder, J.K., Logananthan, G.V., Agbenowski, N., and Grablutz, F. (2003) Prioritizing Water Main Renewal Program, *Pipelines 2003*, 50.

De Silva, D., Moglia, M., Davis, P., and Burn, S. (2002) Condition Assessment and Probabilistic Analysis to Estimate Failure Rates in Buried Pipelines, *Proceedings of ASTT 5th Conference.*

De Silva, D., Moglia, M., Davis, P., and Burn, L.S. (2006) Condition Assessment and Probabilistic Analysis to Estimate Failure Rates in Buried Metallic Pipelines, *Journal of Water Supply: Research and Technology*, AQUA, 55, 175-191.

DeStefano, P. D. and Grivas, D.A. (1998) Method for Estimating Transition Probability in Bridge Deterioration Models, *Journal of Infrastructure Systems*, ASCE, 4(2), 56-62.

Doleac, M. L., Lackey, S. L., and Bratton, G. N. (1980) Prediction of Time-to Failure for Buried Cast Iron Pipe, *Proc. AWWA Annual Conference*, Denver, CO. 31-28.

Ebeling, C.E. (1997) *Reliability and Maintainability Engineering*, McGraw-Hill, New York, ISBN 0-07-018852-1.

Elishakoff, I. (2005) Stochasticity and Safety Factors: Part 1. Random Actual Stress and Deterministic Yield Stress, *Chaos, Solutions and Fractals*, 23:1, 321-331.

Ellingwood, B.R. (2007) Risk-Informed Condition Assessment of Civil Infrastructure: State of Practice and Research Issues, *Structure and Infrastructure Engineering*, 1:1, 7-18.

Esselman, T.C., Mohsen, A.E., and McBrine, W.J. (1998) Structural Condition Monitoring In A Life Cycle Management Program, *Nuclear Engineering and Design*, 181, 163-173.

Faber, M.H. and Stewart, M.G. (2003) Risk Assessment for Civil Engineering Facilities: Critical Overview and Discussion, *Reliability Engineering and System Safety*, 80, 173-184.

Ganoulis, J.G. (1994) *Engineering Risk Analysis of Water Pollution*, VCH, Weinheim, Germany, ISBN 3-527-30050-3.

Herz, R.K. (1996) Aging Processes and Rehabilitation Needs of Drinking Water Distribution Networks, *Journal of Water, SRT-Aqua,* 45, 221-231.

Herz, R. (2002) Software for Strategic Rehabilitation and Investment Planning. *International Conference Computer Aided Rehabilitation of Water Networks CARE-W.* Dresden. 65-84.

Herz, R. and Krug, R. (2000) Sanierungsbedarf und Sanierungsstategien für das Grundwasser. *Gwf-Wasser/Abwasser,* 133 (4), 196-200.

Hochstrate, K. (2000) Sewer Status Assessment by Rehabilitation Priority, Intrinsic Value and Functionality as Basis for Forecast Supported Inspections and Predictive Rehabilitation Planning, *Internationaler Kongress Leitungsbau,* Hamburg, 548-557.

Hörold, S. (1998) Forecasting Sewer Rehabilitation Needs. Evaluation of AQUA Wertmin Software for Service Life and Total Cost Estimation. A Case Study in Trondheim (Norway), *Lehrstuhl Stadtbauwesen*, TU Dresden.

Hoskins, R.P, Brint, A., and Strbac, G. (1998) A Structured Approach To Asset Management within the Electricity Industry, *Utilities Policy*, 7, 221-232.

IPWEA (2006) *International Infrastructure Management Manual*, Australia/NZ Edition, 2006.

Iyyer, N., Sarkar, S., Merrill, R., and Phan, N. (2007) Aircraft Life Management Using Crack Initiation and Crack Growth Models – P-3C Aircraft Experience, *International Journal of Fatigue*, 29, 1584-1607.

Jacobs, P., and Karney, B. (1994) GIS development with application to cast iron water main breakage rate. In *2nd International Conference on Water Pipeline Systems*, BHR Group Ltd, Edinburgh, Scotland.

Jardine, A.K.S, Lin, D., and Banjevic, D. (2006) A Review on Machinery Diagnostics and Prognostics Implementing Condition-Based Maintenance, *Mechanical Systems and Signal Processing*, 20, 1483-1510.

Jarrett, R., Hussain, O., Veevers, A., and Van der Touw, J. (2001) A Review of Asset Management Models Available For Water Pipeline Networks, *Proceedings Of International Conference of Maintenance Societies 2001 (ICOMS-2001)*, Melbourne, Paper 49.

Jarrett, R., Hussain, O. and Van der Touw, J. (2002) Prediction of Failures in Water Mains and Prioritisation for Replacements, *Proceedings of the CSIRO/AWA 2nd Annual Seminar on Asset Management of Pipelines: Risk-Based Performance of Urban Water Pipelines.*

Johnson, V.E. and Albert, J.H. (1999) *Ordinal Data Modeling*, Springer, Amsterdam.

KANEW (2003) available at: www.tu-dresden.de/biwiss/stadtbau/KANEW.html.

Kathula, V.S. (2001) *Structural distress condition modeling for sanitary sewers*, Ph.D. Thesis, Civil of Engineering, Louisiana Tech University, 293p.

Kingston, G.B., Maier, H.R., and Lambert, M.F. (2006) A Probabilistic Method for Assisting Knowledge Extraction from Artificial Neural Networks Used for Hydrological Prediction, *Mathematical and Computer Modeling*, Vol. 4, No. 5-6, 499-512.

Kleiner, Y. (2001) Scheduling Inspection and Renewal of Large Infrastructure Assets, *Journal of Infrastructure Systems*, ASCE, 7, 136-143.

Kleiner, Y. and Rajani, B. (1999) Using Limited Data to Assess Future Needs. *Journal of American Water Works Association*, 91(7), 47-61.

Kleiner, Y. and Rajani, B. (2000) Considering Time Dependent Factors In the Statistical Prediction of Water Main Breaks, *AWWA Infrastructure Conference Proceedings*, Baltimore, Maryland, 134-141.

Kleiner, Y. and Rajani, B. (2001) Comprehensive Review of Structural Deterioration of Water Mains: Statistical Models, *Urban Water Journal*, 3, 131-150.

Kleiner, Y. and Rajani, B. (2002) Forecasting Variations and Trends in Water-Main Breaks, *Journal of Infrastructure Systems*, ASCE, 8(4), 122-131.

Kleiner, Y. and Rajani, B.B. (2004) Quantifying effectiveness of cathodic protection in water mains, *Journal of Infrastructure Systems*, 10(2), 43-51.

Kleiner, Y., Adams, B.J., and Rogers, J.S. (1998) Long-term Planning Methodology for Water Distribution System Rehabilitation, *Water Resources Research*, 34(8), 2039-2051.

Kleiner, Y., Sadiq, R., and Rajani, B. (2004) Modeling Failure Risk in Buried Pipes Using Fuzzy Markov Deterioration Process, *Pipelines*, 7-16.

Kleiner Y., Sadiq R., and Rajani, B. (2006) Modeling the Deterioration of Buried Infrastructure as a Fuzzy Markov Process, *Journal of Water Supply Research and Technology-Aqua*, 55(2), 67-80.

Kleiner, Y., Rajani, B., and Sadiq, R. (2004) Management of Failure Risk in Large-Diameter Buried Pipes using Fuzzy-based Techniques, *4th International Conference on Decision Making in Urban and Civil Engineering*, Porto, Portugal, Oct. 28-30, 2004, pp. 1-11.

Kleiner, Y., Rajani, B., and Wang, S. (2007) Consideration of Static and Dynamic Effects to Plan Water Main Renewal, *Middle East Water 2007, 4th International Exhibition and Conference for Water Technology*, Manama, Bahrain.

Koo, D.-H. and Ariaratnam, S.T. (2006) Innovative Method for Assessment of Underground Sewer Pipe Condition, *Automation in Construction*, 15, 479-488.

Kulkarni, R.B., Golabi, K., and Chuang, J. (1986). *Analytical Techniques for Selection of Repair-or-Replace Options for Cast Iron Gas Piping Systems – Phase I*, Gas Research Institute, PB87-114112, Chicago, IL.

Le Gat, Y., Bremond, B., and Fougerat, B. (1996) Statistical Modeling of The Risk of Abnormalities in the Sewerage Network cannot be Visited. *in: Behavior of COS'96 Articles of Service.* Clermont-Ferrand, 9-10 May *1996.*

LGA (2001) *Infrastructure Asset Depreciation Project, Draft Report Section 1 and 2,* Local Government Association of South Australia.

Li, C.Q., Mackie, R.I., and Lawanwisut, W. (2007) A Risk-Cost Optimized Maintenance Strategy for Corrosion-Affected Concrete Structures, *Computer-Aided Civil and Infrastructure Engineering* 22, 335-346.

Lillie, K., Reed, C., and Rodgers, M.A.R. (2004) *Workshop on Condition Assessment Inspection Devices for Water Transmission Mains*, AwwaRF, U.S.

Lou, Z., Gunaratne, M., Lu, J.J., and Dietrich, B. (2001) Application of Neural Network Model to Forecast Short-Term Pavement Crack Condition: Florida Case Study, *Journal of Infrastructure Systems*, ASCE, Vol. 7(4), 166-171.

Madanat, S. and Ibrahim, W.H.W. (1995) Poisson Regression Models of Infrastructure Transition Probabilities, *Journal of Transportation Engineering*, ASCE, 121(3), 267-272.

Madanat, S., Mishalani, R., and Ibrahim, W.H.W. (1995). Estimation of Infrastructure Transition Probabilities from Condition Rating Data, *Journal of Infrastructure Systems*, ASCE, 1(2), 120-125.

Madryas, C. and Przbyla, B. (1998) Inspection of Pipes as an Element of Operating Municipal Sewerage Networks, *Tunneling & Underground Space Technology*, 13(1), 57-64.

Maier H.R. and Dandy, G.C. (2001). Neural Network Based Modeling of Environmental Variables: A Systematic Approach, *Mathematical and Computer Modeling*, 33(6), 669-682.

Marlow, D. and Burn, S. (2008) Effective use of Condition Assessment within Asset Management, *Journal of the AWWA*, 100, 1.

Marlow, D. and Burn, S. (2007) Capturing Externalities for Optimum Whole of Life Costing, *IQPC Conference – Whole of Life Costing*, Sydney, March, 2007.

Marlow, D. (2007) Best Practice Risk and Asset Management, *National Water 2007*, 14-16 August, Melbourne, Australia.

Marvin, C. (1996) *Aggressive Maintenance Positively Pays Big Dividends to Building Owner*, RCI Interface, Roofing Consultants Institute, July.

Mauch, M. and Madanat, S. (2001) Semiparametric Hazard Rate Models of Reinforced Concrete Bridge Deck Deterioration, *Journal of Infrastructure Systems*, ASCE, 7(2), 49-57.

McDonald, S.E. and Zhao, J.Q. (2001) Condition Assessment and Rehabilitation of Large Sewers, *International Conference on Underground Infrastructure Research*, University of Waterloo, Waterloo, Ontario, June 10-13, 2001, pp. 361-369.

Melchers, R.E. (1996) Towards a probabilistic model for marine corrosion of steel. In Frangopol, D M. & Grigoriu, M. D. (Eds.) *7th Specialty Conference on Probabilistic Mechanics and Structural Reliability,* Worcester, MA, Amer Soc Civil Engineers.

Melchers, R.E. (2005) The Effect of Corrosion on the Structural Reliability of Steel Offshore Structures, *Corrosion Science*, 47, 2391-2410.

Micevski, T., Kuczera, G., and Coombes, P. (2002) Markov Model for Stormwater Pipe Deterioration, *Journal of Infrastructure Systems*, ASCE, 8(2), 49-56.

Mishalani, R.G. and Madanat, S.M. (2002) Computation of Infrastructure Transition Probabilities using Stochastic Duration Models, *Journal of Infrastructure Systems*, ASCE, 8(4), 139-148.

Moglia, M., Burn, S., and Meddings, S. (2006) A Decision Support System for Reducing the Predicted Consequences of Water Reticulation Pipe Failures, *Journal of Information Technology in Construction (ITcon)*, Vol. 11, Special Issue Decision Support Systems for Infrastructure Management, 237-256, http://www.itcon.org/2006/18.

Moglia, M., Davis, P., and Burn, S. (2008) Strong Exploration of a Cast Iron Pipe Failure Model. *Reliability Engineering & System Safety*, 93, 885-896.

Morcous, G., Rivard, H., and Hanna, A.M. (2002) Modeling Bridge Deterioration Using Case-based Reasoning, *Journal of Infrastructure Systems*, ASCE, 8(3), 86-95.

MUV-BW (Ministerium für Umwelt und Verkehr Baden-Württemberg (2000) *Leitfaden für kostenminimierende Instandhaltung von Kanalnetzen.* (Guidelines for cost minimized sewer rehabilitation) Stuttgart. 49 p.

Nelson, W. (1982) *Applied Life Data Analysis*, Wiley, New York.

Newton, L. A. and Vanier, D (2006) *MIIP Report: The State of Canadian Sewers - Analysis of Asset Inventory and Condition*, National Research Council, Institute for Research in Construction NRC-IRC, Ottawa, Canada.

NGSMI (2003) Selection of Technologies for Sewer Rehabilitation and Replacement, *National Guide to Sustainable Municipal Infrastructure*, Issue No. 1.0, March 2003.

NSW (2001) *Asset Maintenance: Strategic Planning*, New South Wales. Government Asset Management Committee, DPWS Report Number 01047, ISBN 0 7347 4134 0, January 2001.

NSW (2006) *Total Asset Management (TAM) Guidelines*, see documents at http://www.treasury.nsw.gov.au/tam/tam-guide, accessed March 2008.

Ofwat (1998) *PR99 Information Requirement H, Asset Inventory and System Performance Submission; Reporting Requirements and Definitions*, Ofwat, UK, August 1998.

Pohls, O. (2001) *The Analysis of Tree Root Blockages in Sewer Lines & their Prevention Methods*. Institute of Land and Food Resources. University of Melbourne.

Rahman, S. and Vanier, D.J. (2004) An Evaluation of Condition Assessment Protocols for Sewer Management, *Municipal Infrastructure Investment Planning (MIIP) Report B-5123.6*, CNRC, July 2004.

Rajani, B. and Makar, J. (2000) A Methodology to Estimate Remaining Service Life of Grey Cast Iron Water Mains, *Can J Civil Engng,* 27, 1259-1272.

Rajani, B., Zhan, C., and Kuraoka, S. (1996) Pipe-soil Interaction Analysis for Jointed Water Mains, *Canadian Geotechnical Journal*, 33(3), June 1996, 393-404.

Rajani, B. and Kleiner, Y. (2001) Comprehensive Review of Structural Deterioration of Water Mains: Physically Based Models, *Urban Water*, 3(3), 151-164.

Rajani, B., Kleiner, Y., and Sadiq, R. (2006) Translation of Pipe Inspection Results Into Condition Ratings Using The Fuzzy Synthetic Evaluation Technique, *Journal of Water Supply Research and Technology-Aqua*, 55, 11-24.

Rajani, B.B. and Kleiner, Y. (2007) Quantifying effectiveness of cathodic protection in water mains: case studies, *Journal of Infrastructure Systems*, 13(1), 1-11.

Rajani, B. and Zhan, C. (1996) On the Estimation of Frost Loads, Canadian *Geotechnical Journal, Revue Canadienne De Geotechnique*, 33(4), 629-641.

Rahman, S. 1997. Probabilistic Fracture Analysis of Cracked Pipes with Circumferential Flaws, *International Journal of Pressure Vessels and Piping*, 70, 223-236.

Randall-Smith, M., Russell, A., and Oliphant, R. (1992) *Guidance manual for the structural condition assessment of trunk mains*, WRc, UK.

Rausand, M. and Øien, K. (1996) The Basic Concepts of Failure Analysis, *Reliability Engineering and Systems Safety*, 53, 73-83.

Ross, M.S. (1972). *Introduction to Probability Models*, McGraw-Hill, New York.

Rossum, J.R. (1969) Prediction of Pitting Rates In Ferrous Metals From Soil Parameters*, Journal of the AWWA*, *61*, 305-310.

Sadiq, R., Kleiner, Y. and Rajani, B. (2003) Forensics of water quality failure in distribution system - a conceptual framework, *Journal of Indian Water Works Association*, 35(4), 267-278.

Sadiq, R., Rajani, B, and Kleiner, Y. (2004) Probabilistic Risk Analysis of Corrosion Associated Failures in Cast Iron Water Mains, *Reliability Engineering and System Safety*, 86, 1-10.

Schneider, J., Gaul, A.J., Neumann, C., Hogräfer, J., Wellßow, W., Schwan, M., and Schnettler, A. (2006), Asset Management Techniques, *International Journal of Electrical Power and Energy Systems,* 28(9), 643-654.

Schuyler, J R (1997) *Decision Analysis in Projects*, Gilmour Drummond Publishing, United Kingdom.

Shamir, U. and Howard, D. (1979) An Analytic Approach to Scheduling Pipe Replacement, *Journal of the AWWA*, May 1979, 248-258.

Smith, M. (1996) *Neural Networks For Statistical Modeling*, International Thompson Computer Press, London.

Speers, A., Burn, S., Hatton MacDonald, D., Young, M., and Syme, G. (2002) Setting and Evaluating Customer Service Standards, *IWA 3rd World water congress: Enviro 2002*, Melbourne, Australia.

Steinbuch, R. (1999) Comparison of Critical Circumferential Through-Wall Crack Lengths In Welds Between Pieces of Straight Pipes to Welds Between Straight Pipes and Bends, With and Without Internal Pressure at Force- and Displacement-Controlled Bending Load, *Nuclear Engineering and Design,* 193, 297-308.

Tabachnick, B. G., and Fidell, L.S. (2001) *Using Multivariate Statistics*, Allyn & Bacon, New York.

Taylor, G. J. (1993) *The Promise of Neural Networks*, Springer-Verlag.

Thoft-Christensen, P. and Baker, M.J. (1982) *Structural Reliability Theory and its Applications*, ed. Anonymous., Berlin: Springer-Verlag. 267.

Tran, D.H., Ng, A.W.M., and Perera, B.J.C. (2007) Neural Network Deterioration Models for Serviceability Condition of Buried Stormwater Pipes, *Engineering Applications of Artificial Intelligence*, 20, 1144-1151.

Tran, H.D. (2008) *Investigation of Deterioration Models for Stormwater Pipe Systems*, PhD Dissertation, Victoria University, Melbourne.

U.S. EPA (2002a) *The Clean Water and Drinking Water Infrastructure Gap Analysis*, United States, Office of Water (4606M), EPA-816-R-02-020.

U.S. EPA (2002b) *SSO Fact Sheet – Asset Management for Sewer Collection Systems*, Office of Wastewater Management, 833-F-02-001, April 2002.

UKWIR (2002) *Capital Maintenance Planning: A Common Framework Volume 1: Overview*, UKWIR Ref 02/RG/05/3, see http://www.ukwir.org/ukwirlibrary/80474 (accessed March 2008).

Val, D.V. and Stewart, M.G. (2005) Decision Analysis for Deteriorating Structures, *Reliability Engineering and System Safety*, 87, 377-385.

Vanier, D.J. (2001) Asset management: "A" to "Z", *APWA International Public Works Congress*, Philadelphia, September, 2001.

Vanier, D.J. and Rahman, S. (2004) A Primer on Municipal Infrastructure Asset Management, *Municipal Infrastructure Investment Planning (MIIP) Report: B-5123.3*, CNRC, May 2004.

Veevers, A. and Hussain, A.Z.M.O. (1999) Reliability Modeling and Asset Management for Distribution Systems, *International Journal of Reliability, Quality and Safety Engineering*, 1999. 6(3), 217-228.

Walski, T.M., and Pelliccia, A. (1982) Economic Analysis of Water Main Breaks, *Journal of the AWWA,* 74(3), 140-147.

Wang, W. and Zhang, W. (2008) An Asset Residual Life Prediction Model Based on Expert Judgments, *European Journal of Operational Research*, 188, 496-505.

Watson, T., Christian, C., Mason, A., and Smith, M. (2001) Maintenance of Water Distribution Systems, *Proceedings of the 36th Annual Conference of the Operational Research Society of New Zealand*, University of Canterbury, New Zealand, 30 November - 1 December, 2001, 57-66.

WERF (2007) *Condition Assessment Strategies and Protocols for Water & Wastewater Utility Assets*, prepared by Marlow et al., WERF, Alexandria, VA, Project 03-CTS-20CO.

Wirahadikusumah, R., Abraham, M. D., and Iseley, T. (2001) Challenging Issues in Modeling Deterioration of Combined Sewers, *Journal of Infrastructure Systems*, ASCE, 7(2), 77-84.

Woodhouse, J. (1999) *Cost Risk Optimisation*, downloaded from http://www.plant-maintenance.com/articles/Costriskop.pdf, accessed March 2008.

WRc (2001) *Sewer Rehabilitation Manual (SRM) 4th Edition, Version 2*, November 2001, see http://www.wrcplc.co.uk/srm/ (accessed March 2008).

Zhao, J.Q. (1998) Trunk Sewers in Canada, *Proceedings of the APWA International Public Works Congress – Innovations in Urban Infrastructure*, Las Vegas, NV. pp. 75-89.

WERF SUBSCRIBERS

WASTEWATER UTILITY

Alabama
Montgomery Water Works & Sanitary Sewer Board

Alaska
Anchorage Water & Wastewater Utility

Arizona
Avondale, City of
Glendale, City of, Utilities Department
Mesa, City of
Peoria, City of
Phoenix Water Services Dept.
Pima County Wastewater Management
Safford, City of
Tempe, City of

Arkansas
Little Rock Wastewater Utility

California
Central Contra Costa Sanitary District
Corona, City of
Crestline Sanitation District
Delta Diablo Sanitation District
Dublin San Ramon Services District
East Bay Dischargers Authority
East Bay Municipal Utility District
El Dorado Irrigation District
Fairfield-Suisun Sewer District
Fresno Department of Public Utilities
Inland Empire Utilities Agency
Irvine Ranch Water District
Las Gallinas Valley Sanitary District
Las Virgenes Municipal Water District
Livermore, City of
Los Angeles, City of
Los Angeles County, Sanitation Districts of
Napa Sanitation District
Novato Sanitary District
Orange County Sanitation District
Palo Alto, City of
Riverside, City of
Sacramento Regional County Sanitation District
San Diego Metropolitan Wastewater Department, City of
San Francisco, City & County of
San Jose, City of
Santa Barbara, City of
Santa Cruz, City of
Santa Rosa, City of
South Bayside System Authority
South Coast Water District
South Orange County Wastewater Authority
Stege Sanitary District
Sunnyvale, City of
Union Sanitary District
West Valley Sanitation District

Colorado
Aurora, City of
Boulder, City of
Greeley, City of
Littleton/Englewood Water Pollution Control Plant
Metro Wastewater Reclamation District, Denver

Connecticut
Greater New Haven WPCA
Stamford, City of

District of Columbia
District of Columbia Water & Sewer Authority

Florida
Broward, County of
Fort Lauderdale, City of
Jacksonville Electric Authority (JEA)
Miami-Dade Water & Sewer Authority
Orange County Utilities Department
Pinellas, County of
Reedy Creek Improvement District
Seminole County Environmental Services
St. Petersburg, City of
Tallahassee, City of
Tampa, City of
Toho Water Authority
West Palm Beach, City of

Georgia
Atlanta Department of Watershed Management
Augusta, City of
Clayton County Water Authority
Cobb County Water System
Columbus Water Works
Fulton County
Gwinnett County Department of Public Utilities
Savannah, City of

Hawaii
Honolulu, City & County of

Idaho
Boise, City of

Illinois
Decatur, Sanitary District of
Greater Peoria Sanitary District
Kankakee River Metropolitan Agency
Metropolitan Water Reclamation District of Greater Chicago
Wheaton Sanitary District

Iowa
Ames, City of
Cedar Rapids Wastewater Facility
Des Moines, City of
Iowa City

Kansas
Johnson County Wastewater
Lenexa, City of
Unified Government of Wyandotte County/ Kansas City, City of

Kentucky
Louisville & Jefferson County Metropolitan Sewer District

Louisiana
Sewerage & Water Board of New Orleans

Maine
Bangor, City of
Portland Water District

Maryland
Anne Arundel County Bureau of Utility Operations
Howard County Bureau of Utilities
Washington Suburban Sanitary Commission

Massachusetts
Boston Water & Sewer Commission
Massachusetts Water Resources Authority (MWRA)
Upper Blackstone Water Pollution Abatement District

Michigan
Ann Arbor, City of
Detroit, City of
Holland Board of Public Works
Saginaw, City of
Wayne County Department of Environment
Wyoming, City of

Minnesota
Rochester, City of
Western Lake Superior Sanitary District

Missouri
Independence, City of
Kansas City Missouri Water Services Department
Little Blue Valley Sewer District
Metropolitan St. Louis Sewer District

Nebraska
Lincoln Public Works and Utilities Department

Nevada
Henderson, City of
Las Vegas, City of
Reno, City of

New Jersey
Bergen County Utilities Authority
Ocean County Utilities Authority
Passaic Valley Sewerage Commissioners

New York
New York City Department of Environmental Protection

North Carolina
Charlotte/Mecklenburg Utilities
Durham, City of
Metropolitan Sewerage District of Buncombe County
Orange Water & Sewer Authority
University of North Carolina, Chapel Hill

Ohio
Akron, City of
Butler County Department of Environmental Services
Columbus, City of
Metropolitan Sewer District of Greater Cincinnati
Montgomery, County of
Northeast Ohio Regional Sewer District
Summit, County of

Oklahoma
Oklahoma City Water & Wastewater Utility Department
Tulsa, City of

Oregon
Albany, City of
Clean Water Services
Eugene, City of
Gresham, City of
Portland, City of Bureau of Environmental Services
Water Environment Services

Pennsylvania
Hemlock Municipal Sewer Cooperative (HMSC)
Philadelphia, City of
University Area Joint Authority

South Carolina
Charleston Water System
Mount Pleasant Waterworks & Sewer Commission
Spartanburg Water

Tennessee
Cleveland Utilities
Murfreesboro Water & Sewer Department
Nashville Metro Water Services

Texas
Austin, City of
Dallas Water Utilities
Denton, City of
El Paso Water Utilities

Fort Worth, City of
Houston, City of
San Antonio Water System
Trinity River Authority

Utah
Salt Lake City Corporation

Virginia
Alexandria Sanitation Authority
Arlington, County of
Fairfax County
Hampton Roads Sanitation District
Hanover, County of
Henrico, County of
Hopewell Regional Wastewater Treatment Facility
Loudoun Water
Prince William County Service Authority
Richmond, City of
Rivanna Water & Sewer Authority

Washington
Everett, City of
King County Department of Natural Resources
Seattle Public Utilities
Sunnyside, Port of
Yakima, City of

Wisconsin
Green Bay Metro Sewerage District
Kenosha Water Utility
Madison Metropolitan Sewerage District
Milwaukee Metropolitan Sewerage District
Racine, City of
Sheboygan Regional Wastewater Treatment
Wausau Water Works

Australia
ACTEW (Ecowise)
South Australian Water Corporation
South East Water Limited
Sydney Water Corporation
Water Corporation of Western Australia

Canada
Edmonton, City of/Edmonton Waste Management Centre of Excellence
Lethbridge, City of
Regina, City of, Saskatchewan
Toronto, City of, Ontario
Winnipeg, City of, Manitoba

New Zealand
Watercare Services Limited

United Kingdom
Anglian Water Services, Ltd.

STORMWATER UTILITY

California
Fresno Metropolitan Flood Control District
Los Angeles, City of, Department of Public Works
Monterey, City of
San Francisco, City & County of
Santa Rosa, City of
Sunnyvale, City of

Colorado
Aurora, City of
Boulder, City of

Florida
Orlando, City of

Iowa
Cedar Rapids Wastewater Facility
Des Moines, City of

Kansas
Overland Park, City of

Kentucky
Louisville & Jefferson County Metropolitan Sewer District

Maine
Portland Water District

North Carolina
Charlotte, City of, Stormwater Services

Pennsylvania
Philadelphia, City of

Tennessee
Chattanooga Stormwater Management

Texas
Harris County Flood Control District, Texas

Washington
Bellevue Utilities Department
Seattle Public Utilities

STATE

Connecticut Department of Environmental Protection
Kansas Department of Health & Environment
New England Interstate Water Pollution Control Commission (NEIWPCC)
Ohio River Valley Sanitation Commission
Urban Drainage & Flood Control District, CO

CORPORATE

ADS Environmental Services
Advanced Data Mining International
AECOM
Alan Plummer & Associates
Alpine Technology Inc.
Aqua-Aerobic Systems Inc.
Aquateam–Norwegian Water Technology Centre A/S
ARCADIS
Associated Engineering
Bernardin Lochmueller & Associates
Black & Veatch
Blue Water Technologies, Inc.
Brown & Caldwell
Burgess & Niple, Ltd.
Burns & McDonnell
CABE Associates Inc.
The Cadmus Group
Camp Dresser & McKee Inc.
Carollo Engineers Inc.
Carpenter Environmental Associates Inc.
CET Engineering Services
CH2M HILL
CRA Infrastructure & Engineering
CONTECH Stormwater Solutions
D&B/Guarino Engineers, LLC
Damon S. Williams Associates, LLC
Ecovation
EMA Inc.
Environmental Operating Solutions, Inc.
Environ/The ADVENT Group, Inc.
Fay, Spofford, & Thorndike Inc.
Freese & Nichols, Inc.
ftn Associates Inc.
Gannett Fleming Inc.
Garden & Associates, Ltd.
Geosyntec Consultants
GHD
Greeley and Hansen LLC
Hazen & Sawyer, P.C.
HDR Engineering Inc.
HNTB Corporation
Hydromantis Inc.
HydroQual Inc.
Infilco Degremont Inc.
Jacques Whitford NAWE, Inc.
Jason Consultants LLC Inc.
Jordan, Jones, & Goulding Inc.
KCI Technologies Inc.
Kelly & Weaver, P.C.
Kennedy/Jenks Consultants
KMK Consultants
Larry Walker Associates
Limno-Tech Inc.
The Low Impact Development Center Inc.
Malcolm Pirnie Inc.
Material Matters
McKim & Creed
MPR Engineering Corporation, Inc.
MWH
NTL Alaska, Inc.
O'Brien & Gere Engineers Inc.
Odor & Corrosion Technology Consultants Inc.
Paradigm Environmental Technologies, Inc.
Parametrix Inc.
Parsons
Post, Buckley, Schuh & Jernigan
Praxair, Inc.
Ring Industrial Group
RMC Water & Environment
Ross & Associates Ltd.
SAIC
Siemens Water Technologies
The Soap & Detergent Association
Smith & Loveless, Inc.
Southeast Environmental Engineering, LLC
Stearns & Wheler, LLC
Stone Environmental Inc.
Stratus Consulting Inc.
Synagro Technologies Inc.
Tetra Tech Inc.
Trojan Technologies Inc.
Trussell Technologies, Inc.
Uni-Bell PVC Pipe Association
URS Corporation
Westin Engineering Inc.
Wright Water Engineers
Zoeller Pump Company

INDUSTRY

American Electric Power
American Water
Anglian Water Services, Ltd.
Chevron Energy Technology
The Coca-Cola Company
Dow Chemical Company
DuPont Company
Eastman Chemical Company
Eli Lilly & Company
Johnson & Johnson
Merck & Company Inc.
Procter & Gamble Company
Suez Environnment
United Utilities North West (UUNW)
United Water Services LLC

Note: List as of 2/21/09

WERF Board of Directors

WERF Research Council

IWA Publishing's authorised EU representative for General Product Safety Regulations is Diane D'Arras, 15 rue Duret, 75116 Paris, France, e-mail: safety@iwap.co.uk.

Printed and bound by CPI Group (UK) Ltd, Croydon, CR0 4YY
12/05/2026
02108879-0015